Handbook of Biochemical Testing

Handbook of Biochemical Testing

Edited by **Oliver Stone**

New York

Published by Callisto Reference,
106 Park Avenue, Suite 200,
New York, NY 10016, USA
www.callistoreference.com

Handbook of Biochemical Testing
Edited by Oliver Stone

International Standard Book Number: 978-1-63239-370-8 (Hardback)

Printed in the United States of America.

Contents

Preface

The purpose of the book is to provide a glimpse into the dynamics and to present opinions and studies of some of the scientists engaged in the development of new ideas in the field from very different standpoints. This book will prove useful to students and researchers owing to its high content quality.

Biochemical testing requires the determination of various parameters and the recognition of the main biological chemical compounds, by the usage of biochemical and molecular devices. The objective of this book is to present various processes and procedures to confine and classify unknown bacteria through the molecular, biochemical differences. This can be done depending upon the characteristic gene sequences. Furthermore, molecular tools involving DNA sequencing and biochemical tools based in enzymatic reactions and proteins reactivity will aid to recognize genetically modified organisms in agriculture, as well as for food preservation and healthcare, and improvement through natural products utilization, vaccination and prophylactic treatments, and drugs testing in medical trials.

At the end, I would like to appreciate all the efforts made by the authors in completing their chapters professionally. I express my deepest gratitude to all of them for contributing to this book by sharing their valuable works. A special thanks to my family and friends for their constant support in this journey.

Editor

Part 1

Applied Environmental Microbiology: Isolation, Characterization and Identification of Bacteria

Biochemical Isolation and Identification of Mycobacteria

Wellman Ribón
Universidad Industrial de Santander
Bucaramanga,
Colombia

1. Introduction

The identification of the species that comprise the *Mycobacterium* genus is one of the best documented examples of scientific development and technology. Approximately 150 species are currently recognized that, besides comprising a genus because of structural and biochemical similarities, exhibit a great variety of characteristics that permit the establishment of differential patterns in order to identify and characterize them, or to establish clusters or complexes. At the present time it is insufficient for the performance of tests that would determine if a given specie belongs to the *Mycobacterium* genus. There is clearly a pressing need to establish its membership in a group or complex, and to determine its specie and genetic characterization so as to permit a comparison of characteristics such as frequency and distribution at a global level.

Within this bacterial genus appear species that, for their public health impact, have been extensively studied; this is in contrast with those that have been recently identified, or of limited frequency, and of which only general characteristics are known.

These species share conventional microbiological methodologies that are based on the classification made by Dr. Runyon in designating them as photochromogens, scotochromogens, and non-chromogens, depending on their capacity, or inability, to synthesize pigments with or without light stimulus. Other characteristics, such as the speed of growth, have been used in differentiation through a wide variety of biochemical tests. The development of molecular biology contributes to the identification and characterization of species in only one test, thus requiring less time, and promotes the process of biosecurity as it does not necessitate a live microorganism beyond the initial steps of each experimental protocol. The advance of the science is evidenced through the adoption of recent methodologies that quantify volatile compounds through the odors of a patient or from certain clinical specimens; the use of nanoparticles and mass spectrometry appear to be the most promising developments for the very near future.

In conclusion, the analysis and documentation demonstrate that we do not yet possess the ideal methodology for the identification of mycobacteria. The most promising outlook for the identification of the species that comprise the *Mycobacterium* genus can be found in the recognition of its biochemical characteristics; the use of conventional methodologies; an appreciation of the techniques of molecular biology in its contribution to science and the mitigation of human suffering, while being reasonable in its use; and an unceasing advancement toward the development and implementation of new methodologies produced through scientific, technical, and social advancement.

2. Phenotypic identification

The phenotypic identification of mycobacteria is based on the characteristics of the culture and biochemical features, and comprises a large variety of tests, some of which are conducted in order to locate the microorganisms into two large groups:

1. Species that comprise the *Mycobacterium tuberculosis* complex: to date, it has been reported that 9 species are included in this complex and, as the species are closely-related genetically, occasionally the phenotypic methodologies are unable to discriminate among some of the species, being able to determine only membership in the *M. tuberculosis* complex (Brosch et al 2002).

    M. tuberculosis, as a causal agent of tuberculosis, is the species with the greatest world-wide distribution. The other species that comprise this complex are: *M. bovis, M. africanum, Bacille Calmette- Guerín (BCG)* vaccine strain, *M. microti, M. cannetti, M. caprae, M. pinnipedii, and M. mungi.*

    The following tests and phenotypic observations identify this complex: velocity of growth; colony morphology; pigment production; niacin test; nitrate reduction; catalase activity; formation of the cord factor; urease test; pyrazinamidase test; growth in the presence of p-nitrobenzoic acid; and, growth in the presence of hydrazide of thiophene-2-carboxylic acid.

2. Non-Tuberculous Mycobacteria (NTM): reports indicate that about 140 species have been documented, with similar phenotypic characteristics; a large number of biochemical tests, in different culture media, are needed for identification and observation of the growth of the microorganism; various molecules are added to the culture to highlight the differential characteristics among the species that comprise the *Mycobacterium* genus. These tests should be conducted in conjunction with tests that identify the *M. tuberculosis* complex in order to guarantee an adequate analysis of identification.

3. *M. leprae*: is a non-cultivable species.

The fundamentals of each test used in the identification of mycobacteria are described as follows:

2.1 Characteristics deduced from the direct observation of a culture medium, or growth inhibition, and microscopic observation

2.1.1 Velocity of growth

Mycobacteria are grouped according to velocity of growth as either slow or rapid growth. mycobacteria that develop colonies, visible by eyesight, in a culture medium in less than 7 days are classified as rapid growth; those requiring more than 7 days in which to form visible colonies are designated as slow growth. The rRNA comprises 80% of the RNA of mycobacteria and one unit of its DNA. The RNA produced depends upon the number of *rrn* operons and the efficiency of their transcription. One or two *rrn* operons are found in mycobacteria, regulated by P1 and P2 promoters. The *rrnA operon*, located in the *murA* gene, is found in all species, including *M. tuberculosis* and *M. leprae*; but, in some species, like those of rapid growth, an additional *rrnB* operon is found, located in the *tyrS* gene (Arnvig et al. 2005, Verma et al 1999).

2.1.2 Colony morphology

Mycobacteria in culture generally appear as two types of colonies: rough or smooth, and with a shiny or opaque aspect (Tasso et al 2003).

2.1.3 Pigment

Some species of mycobacteria, classified as photochromogens, produce carotenoid pigment in the presence of light; the group designated as scotochromogenic develops yellow colonies independent of light stimulus; the non-chromogenic group does not produce pigment; their colonies appear with shades of pale yellow or cream, in the presence of light or in darkness, and the color does not intensify when exposed to light (Bernardelli et al 2007, Belén et al 2007).

2.1.4 Growth in presence of P-nitrobenzoic acid

This test contributes to the differentiation of species of the *M. tuberculosis* complex. P-nitrobenzoic acid inhibits the growth of: *M. tuberculosis*, *M. bovis*, *M. africanum*, and *M. microti (Palomino et al 2007)*.

2.1.5 Growth in the presence of hydrazide of thiophene-thiophene-2-carboxilic acid

Permits the differentiation of *M. tuberculosis* that grows in the presence of the compound as opposed to other species such as *M. bovis*; it is sensitive to the compound when added to the Lowenstein Jensen (LJ) medium. Some species of NMT are positive for this test.

2.1.6 Growth in McConkey agar without violet crystal

Mycobacteria such as *M. fortuitum, M. chelonae, and M. abscessus* have the ability to grow in McConkey agar without violet crystal, as opposed to other mycobacteria such as *M. tuberculosis* and *M. bovis* that cannot grow in this medium.

2.1.7 Growth in 5% sodium chloride

Species of mycobacteria such as *M. fortuitum* show the capacity to grow in a LJ culture medium to which has been added 5% sodium chloride.

2.1.8 Growth in the presence of hydroxylamine

This test plays a very important role in determining the difference in capacity of growth of species such as *M. bovis* in LJ media supplemented with a concentration of 250 mg of hydroxylamine. NTM does not have this capability.

2.1.9 Growth in Sauton picric medium and Sauton agar with 0.2% picric acid

This test is fundamental to differentiate mycobacteria of slow and rapid growth. Mycobacteria of rapid growth, with the exception of M. *chelonae*, have the capacity to grow in this medium. Among mycobacteria of slow growth, M. *simie* is unable to grow in this medium.

2.1.10 Cord formation in acid-alcohol resistant bacilli

This involves demonstrating the capacity to form cords from aggregates of acid-alcohol resistant bacilli in which the longitudinal axis of the bacteria is parallel to the longitudinal axis of the cord (Tasso et al 2003). This characteristic is attributed to glycolipid trehalose 6,6-dimycolate or cord factor, which is composed of molecules of mycolic acids (Palomino et al 2007). The presence of the cord in Ziehl Neelsen (ZN) coloration from rough, non-chromogenic colonies is indicative of the presence of M. *tuberculosis*; on the contrary, the absence of a cord, or the uniform distribution of bacilli, is consistent with NTM (Tasso et al 2003).

2.2 Biochemical test of identification

2.2.1 Niacin test

Niacin is part of the energy metabolism of mycobacteria in redox reactions. All mycobacteria produce niacin, but M. *tuberculosis* accumulates it as a result of the major activity of Nicotinamide Adenine Dinucleotide and the inability to process the resulting Niacine (Palomino et al 2007, Cardoso et al 2004)). The test demonstrates the presence of cyanogen chloride formed through the reaction of chloramine T and potassium thiocyanate in the presence of citric acid. The cyanogen chloride breaks the pyridine ring of niacin, forming the aldehyde gamma-carboxyglutamate that binds with the aromatic amine producing a yellow color.

2.2.2 Nitrate reduction test

Nitrate reductase, an enzyme capable of reducing nitrates to nitrites, appears in the cellular membranes of mycobacteria; the bacteria can utilize this enzyme as a source of nitrogen (Palomino et al 2007). The test detects the presence of nitrate reductase in a medium that contains sodium nitrate. The enzyme reduces nitrate to nitrite that appears through the addition of sulfalinamide and dihydrochloride –N- naphtyl ethylendiamine, forming a complex of diazonium chloride with a fuchsia color (Bernardelli et al 2007).

2.2.3 Catalase test

This is an antioxidant enzyme responsible for eliminating molecules of hydrogen peroxide from the cells that are produced during respiration. The reaction results in the release of water and free oxygen (Palomino et al 2007). Two classes of catalase, thermolabile and thermostable, appear in mycobacteria. In M. *tuberculosis* and M. *bovis* the enzymatic activity is inhibited at 68°C; this contrasts with the remaining species of mycobacteria that maintain enzymatic activity following the increase in temperature (Organización Panamericana de la Salud 2008). Oxygen in the form of bubbles, caused by the activity of the enzyme in a solution of perhydrol 30% and Tween 80 at 10%, appears during the test. The height of the column of bubbles can be measured, thus quantifying the enzymatic activity (Bernardelli et al 2007).

2.2.4 Urease test

The urease enzyme is coded by the *ureABC* genes and is able to hydrolyze urea, forming two molecules of ammonium; this is used by the mycobacteria in the process of biosynthesis (Palomino et al 2007). The determination is made by using a medium containing red phenol; production of ammonium results in alkalization of the medium with a change in color to fuchsia.

2.2.5 Pyrazinamidase test

This is an intracellular enzyme codified by the *pncA* gene which is able to hydrolyze pyrazinamide (PZA) in pyrazinoic acid. Some strains present a mutation in the *pncA* gene that generates resistance to PZA, the principal mechanism of resistance in *M. tuberculosis* to this drug. The method of transport of PZA into *M. tuberculosis* is by passive diffusion, where it is converted into pyrazinoic acid through the action of the pyrazinamidase enzyme. Its usefulness as a test of identification is based on the differentiation of *M. tuberculosis* (positive pyrazinamidase) from other species of the *M. tuberculosis* complex (negative pyrazinamidase), with the exception of *M. canetti* which is also positive (Palomino et al 2007, Zhang et al 2003).

2.2.6 Acid phophatase

The acid phosphatase of some mycobacteria separates the free phenolphthalein from phenolphthalein diphosphate using as a substrate of the reaction the magnesium salt thymophthalein monophosphate. The appearance of a red color is positive for acid phosphatase. The *M. tuberculosis* species is negative in this test.

2.2.7 Hydrolysis of polyoxyethylene mono-oleate

This shows the capability of some species of mycobacteria to separate the oleic acid which is esterified in polyoxyethylene monoleate; it is better known commercially and in the laboratory environment as "Tween 80".

2.2.8 Arylsulfatase test

The test is based on the capability of some species of mycobacteria, through the activity of arylsulfatase which acts on sulfate esters, to release its aryl radical. To confirm the activity of the enzyme a color development system should be attached, using phenolphthalein disulfate potassium in the culture medium, that is hydrolyzed by the enzyme producing free phenolphthalein; in the presence of an alkali, a red color appears when the test is positive. Examples of species positive for this test, after three days, are: *M. fortuitum* and *M. abcesuss*; *M. gastri* presents a positive reaction only after two weeks.

2.2.9 Other phenotypic test

Tellurite reduction, oxygen preference, utilization of carbon sources, iron uptake, B galactosidase. (See figure 1).

3. Molecular identification of mycobacteria

The amazing development of biological molecular methods, in the last years, for identification of mycobacteria were implemented for clinical and research uses. Different molecular approaches developed in research laboratories became speedily in diagnostic test. The reference molecular method for identification of mycobateria is the determination of sequences of 16S ribosomal DNA, due to this molecule is highly conserved. (Kirschner et al 1993). The 16S-23S internal transcribed spacer (ITS) sequencing is a supplement to 16S rRNA gene sequencing for identification of closely related species (Roth et al 1998). Others examples of DNA sequencing assays are, *gyrB*, and *rpoB* sequences. The DNA probes used in conjunction with culture methods was employed in many countries for identification of mycobacteria, examples of this methodologies are, AccuProbe, Line probe Assays, INNO LiPA Mycobactera, GenoType Mycobacterium, and GenoType MTBC. Currently, the most reported methodology for mycobacterial identification is the polymerase chain reaction (PCR) restriction-enzyme analysis, this methods is based on the amplification of a 441.bp fragment of the hsp65 gene by PCR and discriminate between all mycobacterial species (*M, tuberculosis* complex species presents the same patrons) (Telenty 1993, Castro et al 2007, 2010, Torres et al 2010).

Different molecular typing methods developed and implemented as a result of the accelerated development of molecular biology have been useful in various contexts of public health and scientific research, among which we can cite: the monitoring of individual clinical isolates of species belonging to the *M. tuberculosis* complex; studies of the transmission of tuberculosis (TB) in urban areas with low, medium, and high rates of transmission; analysis of recent cases of transmission, or TB cases caused by reactivation, reinfection, or mixed diseases; establishment of relationships between remote geographical areas; the influence of transcontinental movements of large numbers of tourists; determination of the geographical origin of isolates; epidemiological analysis of TB outbreaks in closed areas such as prisons, child-care centers, geriatric homes, and indigenous communities; the establishment of the epidemiological nexus of cases that are the product of contacts with a specific patient; and the identification of the index case through a combination of molecular typing techniques.

As typing techniques, both the spoligotyping and Mycobacterial Interspersed Repetitive Units (MIRU) methodologies are the most commonly used tools; they provide a high degree of reproducibility of results and every day are reported in more publications that attest to their versatility and wide use throughout the world; but it cannot be denied that other methodologies exist, which are described below, that have made great contributions in typing processes.

3.1 Typing methods

The *M. tuberculosis* genome is markedly homogeneous with 99.9% similarity and identity to the 16s rRNA sequence, with *M. africanum* and *M. mungi* being species genetically related to the *M. tuberculosis* complex and presenting few mutations (Niobe-Eyangoh et al 2004, Brosh et al 2002). This characteristic has been employed in the development of methodologies that permit the differentiation of species of this complex. As an example of the contribution of this characteristic in the differential identification of *M. tuberculosis* isolates, the following techniques were initially employed: determination of unusual resistance to drugs; serotyping; multilocus enzyme electrophoresis; biochemical heterogeneity, and typing through the use of phages, with the last being the standard technique in use until the end of the 1980's, but presenting problems such as the low number of phage types identified and the labor intensity associated with this experimental protocol. (Jones et al 1975, 1978, Kalndtri et al 2005, Díaz et al 2003). The genome of the species that comprise the *M. tuberculosis* complex presents occasional recombinations caused by mobile fragments or sequences of DNA known as transposons; these are unstable elements potentially capable of causing rearrangements through transposition, deletion, inversion, and duplication, since they are insertion sequences. These mobile insertion elements are responsible for directing the generation of genetic polymorphisms which are commonly employed for discriminating among different isolates. In the studies of *M. tuberculosis* molecular epidemiology the most frequently used insertion sequence is *IS6110*, which is found exclusively in members of the *M. tuberculosis* complex, presenting between 0 – 25 copies which differ in position and number. The study of this element soon became an epidemiological milestone and a methodology of great clinical contribution (Coros et al 2008, Mostrom et al 2002).

In this context the clinical isolates of the *M. tuberculosis* complex that present the same genetic pattern are defined as groupings and it is believed that these cases could have been caused by a recent infection, being part of the same chain of transmission, as opposed to cases of family origin, nosocomial, reinfections, or reactivations. More recently, with genomic sequencing of various members of the *M. tuberculosis* complex, techniques have been employed such as: Sequencing of single nucleotide polymorphism (SNP) spoligotyping, Fluorescent amplified fragment length polymorphism (FAFLP), and, among others. (National TB Controllers Association 2004, Centers for Diseases Control and Prevention (CDC) 2006 and 2007, Caminero et al 2001, Shamputa et al 2007, Small et al 1994, Kamerbeek et al 1997, Kassama et al 2006, Gibson et al 2008, Kremer et al 2005, Orjuela et al 2010, Hernandez et al 2010).

Currently, the methods of molecular typing of members of the *M. tuberculosis* complex can be grouped as (Mostrom 2002, Narayabab et al 2004, Parra et al 2003, and Diaz 2003):

3.1.1 Genomic methods for DNA studies

Restriction Fragment Length Polymorphism IS6110 (RFLP IS6110)
Polymorphic GC rich sequence (PGRS)
Analysis and pulsed-field gel electrophoresis (PFGE)

3.1.2 Methods based on the amplification of specific sequences of DNA using the PCR

Spoligotyping, Fast ligation mediated PCR (Flip), Double-Repetitive Element (DRE-PCR), MIRU, Ligation Mediated PCR (LM-PCR), Fluorescent Amplified Fragment Length Polymorphism (FAFLP), and Single Nucleotide Polymorphism (SNP). (Steinlein et al 2001,

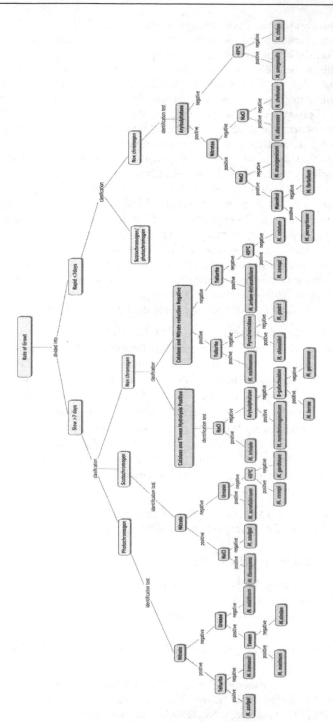

Fig. 1. Flowchart for phenotypic identification of mycobacteria.

Burgos et al 2004, Varela et al 2005, Reisig et al 2005, Kremer et al 2005, Kasama et al 2006, Ashworth et al 2008, Rozo et al 2010, Collins 2011, Li et al 2011, Blaschitz et al 2011).

3.1.3 Typing methods based on genomic DNA

3.1.3.1 RFLP IS6110

Given the current knowledge of the *M. tuberculosis* genome H37Rv, the use of *IS6110* was, until recently, the most frequently reported method for discriminating among bacteriological clinical isolates of members of the *M. tuberculosis* complex; a reduction in its frequency of use is now observed given new methods such as those based on PCR. The RFLP *IS6110* technique has certain disadvantages such as: being labor-intensive; slow; requiring a large quantity of good-quality DNA (1-2 µg); limited ability to discriminate isolates with less than 6 copies of *IS6110* (< 6 bands in the RFLP pattern); and difficulty in comparing results obtained in different laboratories since an international database does not yet exist due to results being reported in patterns and not codes. This molecular marker possesses stability of approximately one to three years in patterns of RFLP *IS6110*, which is related to the number of copies present, since a larger number of copies imply a greater possibility of transposition, especially among isolates of 8 to 12 copies. In general, studies show a high stability in patterns of RFLP *IS6110* (Van et al 1991, Gordon et al 1999, Diel et al 2002, Mostrom et al 2002, Gopaul 2006).

3.1.3.2 PGRS

This method has been of great usefulness in typing clinical isolates of members of the *M. tuberculosis* complex that present a low number of copies of the insertion sequence *IS6110*. These repetitive sequences are present in multiple copies through the mycobacterial genome; approximately 26 to 30 copies of these repetitive sequences are found per chromosome in members of the *M. tuberculosis* complex. It has a discriminating power very similar to that of RFLP, requiring DNA of good quality extracted from cultures of approximately 15 days growth in order to conduct the experimental protocol; several weeks are needed in order to obtain the pattern of each bacterial isolate in electrophoresis gel similar to that obtained through analysis of the *IS6110* sequence. It has a low discrimination in isolates that possess multiple copies of this genetic marker. (Burgos et al 2004, Rozo et al 2010).

3.1.3.3 PFGE

This method possesses a high power of discrimination, requiring large concentrations of DNA with purity and quality; for its implementation it is necessary to make various adaptations in the experimental protocol according to the unique requirements of each laboratory. The introduction of PFGE has had great impact in the study and investigation of the mycobacterial genome, above all in the creation of genomic maps of closely related species such as those that comprise the *M. tuberculosis* complex, thus allowing for the establishment of differences through the identification of multiple rearrangements and of the nonrandom location of insertion elements (Wolfgang et al 1998).

3.1.4 Typing methods based on PCR

3.1.4.1 Spoligotyping

The direct repetitive regions of the "Direct Repeat" sequence in members of the *M. tuberculosis* complex are composed of multiple direct variant repetitive sequences, each of which is comprised of direct repetitions of approximately 36 pairs of bases separated by unique spacer sequences of 35 to 41 pb, generating a large polymorphism which can be used in molecular epidemiological studies for the differentiation of species of the *M. tuberculosis* complex. The methodology of spoligotyping is based on PCR which is targeted on a small DR sequence sandwiched within a spacer region; a total of 94 sequences have been identified, of which 43 are generally used in the genotyping of isolates. All of these can be simultaneously amplified using a single primer set. The presence or absence of these sequences is determined by hybridization with a set of 43 oligonucleotides derived from *M. tuberculosis* H37Rv. This technique has been shown to be useful in the typing of clinical isolates of species of the *M. tuberculosis* complex, especially in those with less than 6 copies of *IS6110*, and in different clinical samples (Bauer 1999, Doroudchi et al 2000, Parra et al 2003).

3.1.4.2 Flip

A rapid method with high reproducibility and great power of discrimination, based on the study of the *IS6110* sequence. In comparison with the RFLP *IS6110* methodology, this technique requires small quantities of DNA (1ng) or crude cell lysates. This methodology has limited discriminating power in isolates with less than six copies of the *IS6110* sequence; it gives results in less time than LM-PCR (Reisig et al 2005).

3.1.4.3 DRE – PCR technique

A methodology with great power of discrimination that consists in the amplification of DNA segments located between the *IS6110* sequence or the polymorphic region rich in guanine cytosine (PGRS). The DRE-PCR method is based on the number of copies and the distance between the repetitive sequences of *IS6110* and PGRS; these distances vary among the different clinical isolates analyzed, and the variations allow a differentiation with respect to the size and number of the amplified fragments of DNA, producing a single pattern band for the different strains of *M. tuberculosis*. It is a method that can be completed using the primary culture of the microorganism and the results can be interpreted after eight hours of work. One of the limitations of this methodology has been the poor power of resolution of the bands procured (Mostron et al 2002).

3.1.4.4 MIRU

A typing methodology based on the varible number tandem repeat (VNTR) or MIRU, with a power of resolution similar to that produced by RFLP *IS6110*, has been useful for typing of members of the *M. tuberculosis* complex. The MIRU are short elements of DNA (40 to 100 pb) found in tandem repetitions and dispersed in intergenic regions of the genome within members of the complex. Given its discriminating power, among the 41 sequences of MIRU that have been described, only 12 or 16 of them are frequently reported as genetic markers. The clinical isolates that are characterized by this genetic marker are designated with a 12-digit code corresponding to the number of repetitions in each MIRU locus, thereby forming the basis of a system that facilitates global comparison among laboratories (Supply et al 2000, 2001, Evans et al 2004, Narayanan et al 2004, Gibson et al 2005).

3.1.4.5 LM-PCR

This is a highly reproducible technique that uses a primer directed toward a specific region of insertion sequence *IS6110*, and a second that is targeted to a linker ligated to restricted genomic DNA. This methodology requires small quantities of DNA and has a high power of discrimination (Prodhom et al 1997, Kremer et al 2005).

3.1.4.6 FAFLP

The FAFLP is a complementary technique for the typing of *M. tuberculosis*, principally in clinical isolates with less than 6 copies of *IS6110*. It has the limitation of requiring the extraction of DNA from the microorganism in growth; subsequently, this is digested with EcoRI and Msel restriction enzymes, although some studies report using BamHI and Mspl (Mortimer 2001, Sims et al 2002).

3.1.4.7 SNP

This methodology is based on the premise that MicroRNAs (miRNA) are thought to play important roles in the pathogenesis of diseases. SNPs within miRNAs can change their characteristics via altering their target selection and/or expression, resulting in functional and/or phenotypic changes. This methodology has produced results similar to those obtained from RFLP *IS6110* and spoligotyping. This technique generates a binary code that permits the creation of a comparative database of different studies (Li et al 2011).

3.1.4.8 Other methodologies

Other methodologies have been developed for the identification and study of mycobacteria such as "in house" PCR (Puerto et al 2007), but statistics have not been reported that would permit an extensive evaluation of these methodologies. There is a large variation among the reported studies which precludes any conclusions regarding their contributions.

4. Recent promising technologies for the detection of *M. tuberculosis*

4.1 Urinary antigen detection

This method is based on the direct detection of lipoarabinomannan (LAM) in urine, when the mycobacteria are metabolically active; accordingly, this glycolipid is found in the urine of patients with active turberculosis. This antigen can be detected through an ELISA test and the determination presents high ranges of sensitivity (from 38 to 51%) and of specificity, close to 89% (Daley et al 2009, Lawn et al 2009, Mutetwa et al 2009, Reither et al 2009, Deng et al 2011).

4.2 Volatile markers

Through chromatographic techniques, various molecules or organic compounds have been identified in human clinical samples with the objective of showing their potential use as markers in the surveillance of illnesses. The volatile organic compounds (VOCs) patterns identified in the *M. tuberculosis* specie through sputum samples allow, through the use of an electronic nose based 14-sensor conducting polymer array, the performance of in vitro and in situ studies. Reports indicate that this methodology identifies 100% of positive *M. tuberculosis* cultures from among others and were able to discriminate between sputum containing either *M. tuberculosis* alone, or a mixed infection. The technique is commonly referred to as "breath testing," with the great advantage of not being an invasive procedure; in a small study it distinguished between patients with a positive sputum culture from those with a negative sputum culture (Pavlou et al 2000, 2004, Buszewski et al 2007, Phillips et al 2007).

4.3 Bead-based methods

This method is based on the application of monoclonal antibodies to nano magnetic beads in order to identify the bacillus. Microsciences Medtech Ltd. (London, United Kingdom) developed a kit of beads coated with a chemical ligand that binds to mycobacteria present in the sputum of patients; the linkage is evident after coloring the preparation and observing it through a fluorescent microscope. Alternative diagnostic techniques have been developed using nanoparticles coupled to PCR, while other methodologies employ monoclonal anti-BCG antibodies. This methodology can be used for detection and identification of *M. tuberculosis* (Kluge et al 2008, Lee et al 2009).

4.4 Simplified smart flow cytometry (S-FC)

Following the use by many researchers of flow cytometry to count CD4 T cells, the S-FC has been recently proposed for arriving at a rapid diagnosis of active TB among Human Immunodeficiency Virus (HIV) negative patients and those with the TB/HIV coinfection. The S-FC is considered to be of great use when a sputum sample cannot be obtained, or among patients with negative sputum, with the recommendation that it be combined with microbiological procedures (Castiblanco et al 2006, Breen et al 2007, Streitz et al 2007, Janossy et al 2008).

4.5 Broad nucleic acid amplification-mass spectrometry

The integration of advances in various technologies such as the technique of analysis and separation of masses, bioinformatics, and ionization techniques linked to mass spectrometry, has facilitated their application in the identification and characterization of pathogens. The successful identification of *M. tuberculosis* complex through the pattern of its mycolic acids has been achieved through the use of Electrospray ionization-tandem mass spectrometry (ESI-MS). Other similar methodologies developed by Abbott Laboratories include the IBIS T5000 which is based upon the amplification of genome sequences of the pathogen linked to mass spectrometry (Ecker et al 2008, 2009, Ho et al 2010, Eshoo et al 2010, Grant et al 2010).

5. Detection of drug susceptibility testing

The increase in TB cases has been accompanied by a growth in the number of patients with isolates of resistant *M. tuberculosis*, thereby confirming TB as a grave, world-wide public health problem. The situation becomes even more complex with the association of *M. tuberculosis* resistant bacilli with HIV, making control more difficult (Bentwich et al 2000, Castiblanco et al 2006, Lemus et al 2004, and World Health Organization (WHO) 2009).

Multidrug-resistant TB (MDR TB), defined as a combined resistance to two first-line medications, isoniazid (INH) and rifampicin (RIF), attained a global level of 500,000 cases in the year 2007 and is expected to continue to increase. According to WHO, 85% of these cases are concentrated in 27 countries The clinical significance of resistance has been defined as the in vitro growth of the microorganism in the presence of the critical concentration of the drug used in treatment, which should be equal or superior to 1% of the growth of the microorganism in the absence of the drug. The critical concentration of a drug is the concentration that inhibits the growth of most wild type within a population without affecting the growth of resistant cells that might be present (Kent et al 1985, Woods et al 2007).

The problem of *M. tuberculosis* resistance appears when the resistant mutants that naturally occur in the microbial population are selected as result of inadequate administration of treatment (irregular administration of dosages, poor absorption, or an inadequate treatment plan) (Parson et al 2004).

When a diagnosis is made of a type of resistance to first-line medications such as INH, RIF, ethambutol (EMB), streptomycin (SM), and PZA, the WHO recommends treatment plans that should be adjusted based upon the pattern of resistance identified in the microorganism and each patient's particular state of health. Second-line medications include: aminoglycosides, cycloserine, capreomycin, fluoroquinolones, and para aminosalicylic acid. These treatment plans are prolonged in comparison with those of the first-line, with increased costs and toxicity, resulting in reduced patient compliance with treatment. When second-line treatment plans are not adequately administered, extensively drug-resistant TB (TB XDR) will probably appear, which is defined as TB MDR with greater resistance to any fluoroquinolone and to at least one of the three injectable medications (amikacin, kanamycin, or capreomycin) (Zumia et al 2001,Gillespie 2002, Raviglione et al 2007, 2004, 2003, 2006, 2007, 2008).

The treatment of TB, compared with that used to cure other infectious diseases, has some differences due to the particular physiological characteristics of *M. tuberculosis,* such as generation time and the capacity to enter into long periods of latency with minimal metabolic activity. In the process that arises from the interaction between the bacteria and the host's defenses, populations of bacilli are identified that are differentiated based upon location, metabolic activity, or velocity of multiplication. Characteristics, therefore, vary depending upon location. For example, when located within pulmonary cavities, they actively multiply; as opposed to being situated in the interior of macrophages where reduced pH and oxygen induce a physiology approximating a state of latency; or, if found within the site of necrosis, they are occasionally able to replicate. Due to the characteristics of the bacillus and the interaction with the host, various effective drugs are used, as appropriate, in the treatment of TB for each of the bacterial populations. Another fundamental aspect in the selection and study of drugs used in the treatment of TB involves the study of the cell structure of the microorganism, in which the most studied has been the bacterial cell wall complex, being hydrophobic, with little permeability, thereby precluding the entry of many therapeutic alternatives (Wayne 1974, Guillespi 2002, Coll 2003, Parson et al 2004, De Rossi et al 2006).

INH,RIF, SM, EMB and PZA are first-line medications for their efficacy in bactericidal or bacteriostatic activity, having less adverse effects than other antimycobacterial medications, as opposed to second-line drugs such as aminoglycosides, polypeptides, fluoroquinolones, ethionamide, cycloserine, and para aminosalicylic acid.

Some characteristics of first- and second-line medications used in the treatment of TB are described below in table 1:

Drug	Nature of compound	Administration way	Mechanism of action	Daily dose	Microbial activity	Susceptible population	Pharmacological interactions	Side effects	7H10 solid medium	LJ solid medium	MGIT	MB/BacT Alert	BACTEC 460
INH	Synthetic	Oral	Inhibition of mycolic acid synthesis	5mg/kg	Bactericidal	Active and dormant bacilli	Fembrina	Hepatotoxic; peripheral neuropathy, seizures	0.2; 1.0	0.2	0.1; 0.4	1.0	0.1; 0.4
RIF	Semisynthetic	Oral	Inhibition of transcription by interacting with RNA polymerase	10mg/kg	Bactericidal	All population	Inhibe anticonceptivos orales, Quinidina	Hepatotoxic; gastric intolerance, hypersensitivity	1.0	40.0	1.0	1.0	2.0
EMB	Synthetic	Oral	Affects cell wall biosynthesis	20mg/kg	Bacteriostatic	Active bacilli	ND	Optic neuritis, alopecia, hypersensitivity	5.0; 10.0	2.0	5.0; 7.5	2.5	2.5; 7.5
PZA	Synthetic	Oral	Affects the membrane potential for bioaccumulation	25mg/kg	Bactericidal	Persistant and dormant bacilli	ND	Arthritis, dermatitis, hepatotoxicity, gastrointestinal disorders	NA	NA	100	ND	100
SM	Natural (Streptomyces griseus)	Parenteral	Blocking RNA translation	15mg/kg	Bactericidal	Active bacilli	Neuromuscular bloking drugs, cross resistance with capreomycin	Nephrotoxicity, ototoxicity	2; 10	4.0	1.0; 4.0	1.0	2.0; 6.0
Amikacin	Semisynthetic	Parenteral	Inhibition of protein synthesis	15mg/kg	Bacteriostatic	Active bacilli	Neuromuscular bloking drugs	Lesion of cranial pair VIII, nefrotoxicity	4.0	ND	1.5	ND	1.0
Capreomycin	Natural (Streptomyces capreolus)	Parenteral	Inhibition of protein synthesis	15-30mg/kg until 1g	Bacteriostatic	Active bacilli	Neuromuscular bloking drugs	Lesion of cranial pair VIII, nefrotoxicity	10.0	40	2.5-3.0	ND	1.25
Ciprofloxacin	Synthetic molecule derived of carbonic acid	Oral and parenteral	Direct inhibition of DNA gyrase	150mg/kg until 12g	Bacteriostatic	Active bacilli	Cross-resistance with other quinolones	Gastrointestinal disorder; CNS; hypersensitivity	2.0	2.0	1.0	ND	2.0
Ethionamide	Synthetic	Oral	Inhibition of mycolic acid synthesis	15-30mg/kg until 1g	Bacteriostatic	Active and dormant bacilli	ND	Gastrointestinal disorder; hepatotoxic	5.0	40	5.0	ND	1.25
Kanamycin	Natural (Streptomyces kanamyceticus)	Parenteral	Inhibition of protein synthesis	15-30mg/kg until 1g	Bactericidal	Active bacilli	Neuromuscular bloking drugs	Lesion of cranial pair VIII, nefrotoxicity	5.0	30	ND	ND	5.0
Levofloxacin	Synthetic molecule derived of carbonic acid	Oral and parenteral	Direct inhibition of DNA gyrase	500mg	Bactericidal	Active and dormant bacilli	Cross-resistance with other quinolones	Tendinitis, neurotoxicity, intestinal disorders, rash, photosensitivity	2.0	2.0	1.5-2.0	ND	ND
Cycloserine	Natural (Streptomyces garyphalus)	Oral	Inhibition of essential pyridoxal 5' phosphate (PLP)-dependent enzymes	10-20mg/kg until 1g	Bacteriostatic and bactericidal	Active and dormant bacilli	Alcohol	Neurotoxicity, psychosis, seizures, depression	25-30	40	ND	ND	≤50
Ofloxacin	Synthetic molecule derived of carbonic acid	Oral	Direct inhibition of DNA gyrase	600-800mg	Bactericidal	Active bacilli	Cross-resistance with other quinolones	Gastrointestinal disorder; CNS; hypersensitivity	2.0	2.0	2.0	ND	2.0
Para-aminosalicylic acid	Synthetic molecule derive from salicilic acid	Oral	Antagonist in the synthesis of folate from all mycobacteria	200mg/kg	Bactericidal	Active bacilli	ND	Gastrointestinal disorder; hepatotoxic	2.0	1.0	ND	ND	4.0

First line: INH, RIF, EMB, PZA, SM
Second line: Amikacin, Capreomycin, Ciprofloxacin, Ethionamide, Kanamycin, Levofloxacin, Cycloserine, Ofloxacin, Para-aminosalicylic acid

References: Svensson et al 1982, Pfyffer et al 1999, Caminero 2003, Coll 2003, Cohenet al 2003, Zhan 2005, CDC 2006, Abate et al 2007, Boldú et al 2007, Loether et al 2010, Cremades et al 2011, Pholwat et al 2011, pholwat et al 2011.

Table 1.

M. tuberculosis acquires resistance to anti-TB agents through mutations in the genome due to such causes as previous exposure to some drug or rearrangements in restricted regions of the DNA of the bacillus; this generally produces resistance of the microorganism to only one medication. No plasmids or transposable elements are involved in this process. (Musser 1995, Quros et al 2001).

Determining the probability that *M. tuberculosis* will develop spontaneous resistance to two drugs used in treatment can be established through the sum of probabilities; for example, a probability exists of 1×10^6-1×10^8 of a combined resistance to INH and RIF, being equal to a total probability of 1×10^{14}. In the process of the development of TB, the total number of possible bacilli in the patient, including cases of chronic cavitary disease, seldom results in this quantity of bacilli (10^{14}). Consequently, the natural development of bacilli resistant to two medications is infrequent; therefore, it is assumed that in the majority of cases, this type of resistance is due to previous exposure to the medications (Long 2000).

Some data suggest that about 17% of new TB cases in a zone of high incidence evidence multiple infections, causing discordance and difficulty in the interpretation of tests and in patient management. The expansion of isolates of resistant *M. tuberculosis* is a concern and creates urgency in the implementation of effective diagnostic measures and control, focusing hopes on the development of a new vaccine, diagnostic methods, and effective alternative therapies (Zumia et al 2001, Richardson et al 2002, Gandhi et al 2006, Palomino et al 2007, WHO 2006, Raizada et al 2009).

5.1 Diagnostic methods of susceptibility

A great variety of methodologies exist for determining the susceptibility of *M. tuberculosis* to medications, among which include standard tests showing the growth of bacillus, such as the method of multiple proportions and BACTEC MIGT 960. Other methods of great impact, and experiencing growing implementation at a global level, are those based on the amplification of nucleic acids through PCR, designated as genotypic methods (Wilson M. et al 2011). Some of these include microarrays of DNA, and solid- phase hybridization, among others. The aforementioned methodologies possess excellent percentages of sensitivity and specificity; however, they vary in the time required to obtain results, cost, and technical complexity, thus complicating their large-scale use, especially in countries with limited resources that employ conventional methods. Rapid phenotypic techniques have recently been endorsed by WHO, and they constitute good solutions for a fast and timely diagnosis of susceptibility compared with standard methodologies, and with the cost and infrastructure required for molecular techniques (Canettii et al 1963, 1969, Siddiqi et al 1981, Telenti et al 1993, Abate et al 1998, Palomino et al 2002, 2005, Mitchison 2005,Lakshmi et al 2006).

5.1.1 Phenotypic methods

Phenotypic methods are extensively used and a wide variety of studies are available that establish their principal advantages and disadvantages. They begin with the cultivation of isolates of *M. tuberculosis* obtained from clinical samples processed in an egg- or liquid-based culture medium, some of which can include antibiotics, which are used for direct determination (observation of the microorganism) or indirect (for example, through color reactions) of the growth of the microorganism and subsequent identification of the resistance of the microorganism through its growth in the presence of anti-TB drugs.

5.1.1.1 Mycobacteria Growth indicator Tube (MGIT BACTEC)

Although various conventional phenotypic methods have been employed in the study of the susceptibility of *M. tuberculosis*, the most frequently reported, using solid media, are: the multiple proportions method; the ratio of resistance and absolute concentrations; and the standard, automated radiometric BACTEC method that, as opposed to the others, uses a culture in liquid medium and the tube growth indicator method. Modifications to the BACTEC system have been adopted in order to avoid the use of radioactive material; these have consisted of using the principle of detecting oxygen consumption and recording the emission of fluorescent signals as a product of the reaction in each test, which is detected and controlled from the interior of the equipment coupled to software algorithms. The system interprets as resistant all cultures with growth equal or greater than that of the tube used as a control in the test (Pffyer et al 1997, 1999, Heifets et al 2000, Lin et al 2009, Yu et al 2011).

5.1.1.2 The Trek Diagnostic System Inc

This method measures the growth of mycobacteria through changes of pressure that occur in the space between the liquid medium and the cap of the tube where the reaction occurs. Results are obtained in approximately three days (Woods et al 2007).

5.1.1.3 The MB/BacT Alert instrument (bioMérieux)

This system emits results from an isolate of *M. tuberculosis* when the bottle that contains the microorganism and the drug is positive before the control tube. The result is obtained through measurement of carbon dioxide released to the medium, detected by a sensor located at the bottom of the bottle used in the test (Siddiqi et al 2006, Woods et al 2007, Lin et al 2009).

5.1.1.4 The multiple proportions method

This method determines the number of mutants resistant to each drug, establishing a relation between the numbers of bacilli that grow in a solid medium with antibiotic as compared with the number of bacilli that grow in a medium without antibiotic. The methodology used with solid medium evaluates the susceptibility of *M. tuberculosis* to first- and second-line medications. Frequently, the method employed uses the LJ culture medium and results are obtained in 21 days, or before, if there are indications of growth (Cannetti et al 1963, 1969, Heifets 2000, WHO 2004, Fisher 2002, Clinical and Laboratory Standards Institute 2003, Woods et al 2007, Pfyffer 1997, 2007).

Currently, BACTEC MGIT (version 960 or 320) and the multiple proportions method are considered as "gold standard" for the diagnosis of TB susceptibility to first- and second-line medications (Pino et al 1998).

As mentioned previously, phenotypic methods of relatively recent implementation now exist, and are noted for having succeeded in establishing a group of non-conventional phenotypic techniques that have been postulated as good diagnostic alternatives for the susceptibility of *M. tuberculosis* depending upon their performance given the particular situations in which they are used.

5.1.1.5 The diagnosis of TB and drug resistance with mycobacteriophages

This group of methodologies offers rapid results, low cost, and uses only viable cells, thereby differing from molecular methodologies. Two types of tests that use mycobacteriophages have been documented: the phage amplified biological assay (phaB) and the luciferase reported phages (LRPs). Both methodologies study the capacity of a clinical isolate of *M. tuberculosis* to harbor an infectious mycobacteriophage and to permit its replication (Albert 2001, Butt et al 2004, Pai et al 2005, and Kalantri et al 2005).

5.1.1.5.1 The PhaB

The PhaB determines the protection and amplification of a phage by mycobacteria in clinical specimens (Eltringham et al 1999). The methodology consists of:

1. Adding phages to the decontaminated clinical sample in order to infect the bacilli present in the samples;
2. Adding a viricidal solution (ammonium ferrous sulfates) that destroys the non-infective phages.
3. Initiating a lytic cycle of the bacilli for the replication of phages; this phase terminates with the release of the phages to the Medium.
4. Released phages added to a cell sensor (*M. smegmatis*) to produce plaques; the formation of plaques occurs if the sample is positive (Albert et al 2004).

5.1.1.5.2 LRPs

The LRPs are phages harboring the firefly luciferase (*flux*) gene, which produces visible light when expressed in the presence of lucifering and cellular ATP (Hazbon 2004). The LRPs behave as molecular vectors of the *flux* gene that encodes for firefly luciferase in the interior of *M. tuberculosis*. Once this genetic information has been incorporated within the bacterial chromosome, it is expressed, and there is availability of ATP and luciferin, photon emissions are produced that are detected by a photographic film in a luminometer. The registering of light emitted indicates the viability of the bacilli (Riska et al 1997, 1998, 1999, Carriere et al 1997, 2003, Bardarov et al 2003).

5.1.1.6 The E-Test

The E-Test determines the pattern of susceptibility of *M. tuberculosis* through growth inhibition halos around strips impregnated with known concentrations of different drugs. Solid medium is used in Middlebrook 7H10 agar plates; the microorganism to be studied should be viable and concentrated in the solution that will be used for inoculation of the medium (Heifets et al 1999, Djiba et al 2004).

5.1.1.7 The colorimetric methods

These methodologies are simple and coupled to oxidation-reduction reactions that employ different indicators. Alamar blue salt is used in the case of REMA and MABA, and bromide 3 (4,5-dimethylthiazol-2-i)-2,5-diphenyltetrazolium, or MTT in the case of TEMA. Fundamentally, the result is obtained through observing a change of color, proportional to the number of viable mycobacteria in the test. TEMA was implemented in 1983 for the study of cellular proliferation and was subsequently utilized to evaluate the viability of microorganisms such as *Staphylococcus aureus* and *Listeria monocytogenes*; it has been extensively used in the study of *M. tuberculosis* resistance. The resulting color in this methodology changes from yellow to purple from the production of insoluble crystals of formazan that can be quantified by spectrophotometry, or a qualitative interpretation can be made by observing the change of color. The methodology provides results within 7 to 9 days in the model proposed by Abate and Mishana for the rapid detection of RIF resistance. The advances of this methodology have included the study of clinical samples applied directly to the test, reporting good values of sensitivity and specificity, close to 100%, for sputum samples. This test permits the determination of the susceptibility of *M. tuberculosis* to first- and second-line drugs in establishing the diagnosis of MDR and XDR TB (Mossmann 1983, Abate et al 1998, Palomino et al 1999, 2002, Morcillo et al 2004, WoldeMeskel et al 2005, Montoro et al 2005, Palomino et al 2007, CDC 2007). The REMA test employs resazurin as an indicator, with an increase in the production of NADPH/NADP, FADH/FAD, FMNH/FMN, and NADH/NAD during multiplication of the bacilli to be studied. The viability of the microorganism is evidenced by the change from blue to pink in the indicator color. This test shows values of sensitivity and specificity exceeding 90% in the evaluation of drugs such as RIF and INH, as well as for second-line drugs. Some reports indicate difficulty in the interpretation of results due to intermediate tones in the change of indicator color (BioSource International, O´brien et al 2000, Maeda et al 2001,Palomino et al 2002, Martin et al 2003, Montoro et al 2005, Rivoire et al 2007).

5.1.1.8 Microscopic-observation drug-susceptibility (MODS)

The MODS are used in various countries with TB MDR problems and permit the rapid observation of the growth of *M. tuberculosis* in culture media with the different first- and second-line drugs employed in the treatment of TB. One of those methods is MODS, that uses an inverted microscope to view the growth of the microorganism. Other methods include a thin layer of agar (TLA) and the HSTB agar method, that use special, transparent, solid culture media, permitting the early observation of the growth of the microorganism with the aid of a conventional microscope (Caviedes et al 2000, Heifets et al 2003, Mejia et al 2004, Robledo et al 2008).

5.1.1.9 The Nitrate Reductase Assay (NRA) or Griess Method

The NRA is based upon the characteristic expressed by mycobacteria to reduce nitrate to nitrite, is a rapid method used in the TB control programs of various countries for the evaluation of the susceptibility of *M. tuberculosis*; this is coupled to processes that permit observation of color change. The test employs the LJ solid medium (Angeby et al 2002, lemus et al 2006, Palomino et al 2007).

5.1.2 Genotypic methods

The genotypic molecular methods for the diagnosis of the susceptibility of *M. tuberculosis* to drugs used in the treatment of TB are based on genetic determinants of resistance, requiring for its development the amplification of specific segments of DNA, and the identification of point mutations in the amplified products, thereby permitting the detection of resistance to medications (Cockerill et al 1999, Telenti et al 1993, Palomino 2005, Fluit et al 2011). Compared to phenotypic methods, these methodologies have the advantage of not requiring the previous growth of the microorganism, with results being obtained in a short time (24 to 48 hours). The following is a description of some methods:

5.1.2.1 DNA sequencing

This methodology is the best and most accurate method for the detection of previously known mutations and for the identification of new mutations. It is the "gold standard" method. It is very effective for detecting mutations to RIF in *M. tuberculosis* by amplification of a single target of study in contrast to studies of various genes involved in resistance.

5.1.2.2 DNA microarrays

This methodology is based on hybridization and allow for the simultaneous analysis of a large quantity of gene segments involved in resistance. The segments amplified by PCR are marked with a fluorescent substance and hybridized with probes attached to a solid phase. Consistent results have been reported for this methodology in the study of resistance to RIF (*rpoB* gene) and genes implicated in resistance to other drugs such as: *KatG, inhA, rpsL*, and *gyrA*. The wide-scale implementation of this methodology has been limited due to the required technical expertise of the operator and the sophisticated equipment. The QIAplex test (Qiagen) system has been recently developed which detects 24 mutations in the *katG, inhA, rpoB, rrs, rpsL*, and *embB* genes (Gegia et al 2008).

5.1.2.3 The Line Probes Assay (LPA)

This method is employed to identify *M. tuberculosis* and determine resistance to RIF, identifying mutations in a specific region of the *rpoB* gene through amplification of this segment using PCR-based reverse hybridization, and using biotin in the test. These labeled products of PCR hybridize with immobilized probes coupling to a color development system that permits the subsequent analysis of resulting color patterns. This test detects four of the most frequent mutations in the *rpoB* gene arguing that 75% of the clinical isolates resistant to RIF have one of these four mutations. The method can be developed from clinical samples with results obtained in 48 hours or less (De Beenhouwer et al 1995, Somoskovi et al 2003, Ramaswamy et al 1998, WHO 2007, Ando et al 2011).

5.1.2.4 Single-strand conformation polymorphism (SSCP)

This methodology detects specific mutations in the region of DNA under study, through the difference in patterns of movement of polyacrylamide gels. The region of the gene involved in resistance should be amplified by PCR and then denatured until two completely separated DNA strands are obtained. These denatured strands migrate by electrophoresis permitting the detection of mutated sequences by the migration pattern, as compared with that which has been established for the wild strain. This methodology has been employed in studies of resistance to RIF; it has not been practical for the investigation of other drugs as it requires the study of various genes involved in resistance to those medications (Kim et al 1997).

5.1.2.5 Real – time PCR or molecular beacons

This methodology is reported as the most rapid methodology among those most frequently used. This technique includes the use of a probe that is designed in the form of a loop that possesses a sequence that is complementary with the sequence of the gene under study, and with the appearance of a fluorescent moiety in one of its extremities. When the probe locates the complementary sequence, hybridization occurs with the emission of fluorescence; the signal is detected and monitored by the associated system. This methodology is highly sensitive and specific; only one change in the sequence under study prevents hybridization, allowing the identification of point mutations. This is a closed system; therefore, minimizing the possibilities of amplification of contaminants. For the study of resistance to RIF, a set of five beacons has been designed, which includes the study of the entire *rpoB* region in only one test. Results are provided in three hours using this methodology, detecting a minimum of two bacilli present in the sample. The test includes the study of genes such as *katG*, *inhA*, *oxyR-ahpC*, and *kasA* (Cockeril 1999, Fluid et al 2001, Piatek et al 1998, 2000, Torres et al 2000, De Viedman 2003).

5.1.2.6 Fluorescence resonance energy transfer (FRET) probes

The FRET is a widely-studied, well-established methodology referred to as "light cycler probes." It is highly specific in identifying mutations in real-time PCR tests. The test permits the detection of RIF and INH resistance in less than two hours with the advantage of facilitating the study of large gene segments involved in resistance (Garcia et al 2002, Torres et al 2003).

5.1.2.7 Other methodologies

Other methodologies have been developed for the study of resistance to drugs used in the treatment of TB such as "in house" PCR, but statistics have not been reported that would permit an extensive evaluation of these methodologies. There is a large variation among the reported studies which precludes any conclusions regarding their contributions.

5.1.2.8 Amplification refractory mutation system (ARMS)

This methodology uses a set of primers, one of which should hybridize with the mutation site; in the absence of this, there is not a place for the amplification of the gene involved in resistance; therefore, the band of the expected product is not observed in the electrophoresis gel. This methodology is used for the study of mutations in the *rpoB*, *katG* y, and *embB* genes with good results (Mokrousov et al 2002, Fan et al 2003).

5.1.2.9 Branch migration inhibition (BMI)

The BMI requires two reactions from PCR, one using DNA from a known strain, and the other using DNA which is the object of study, using a primer labeled with digoxigenin or biotin. This test has been used to study resistance to RIF and PZA; a disadvantage is the identification of silent mutations that conflict with findings reported by phenotypic methods (Lishanski et al 2000, Lyiu et al 2000).

6. Conclusion

The implications of the processes in which mycobacteria participate are diverse and documented but there is a lack of knowledge concerning many of their characteristics. Health, environment, industry, and research are involved with these mycobacterial species in order to advance or improve their processes, verify their findings, or confirm their suspicions. In each of these fields the principal objectives are the detection, identification, and characterization of the species of the *Mycobacterium* genus; only the knowledge of its biological characteristics together with the availability of methods for its study can help us to achieve a complete documentation. The disciplined, periodic and timely revision of the biology, tendencies, innovations, and discoveries regarding this microbial genus should prove to be the best allies of academics, researchers and the general community in addressing any situation that involves mycobacteria in its processes.

7. References

Abate, G., R. Mishana, and H. Miörner. (1998). Evaluation of a colorimetric assay based on 3-(4,5-dimethylthiazol-2-yl)-2,5- diphenyl tetrazolium bromide (MTT) for rapid detection of RIF resistance in *M. tuberculosis*. *Int J Tuberc Lung Dis;* 2(12):1011–6.

Abbate E.H., Palmero D.J., Castagnino J., Cufre M., Doval A., Estevan R., et al.(2007). Tratamiento de la tuberculosis: Guía práctica elaborada por la Sección Tuberculosis, Asociación Argentina de Medicina Respiratoria. Medicina (B. Aires);67(3): 295-305.

Albert, H., A. Trollip, T. Seaman, and R. J. Mole. (2004). Simple, phagebased (FASTPplaque) technology to determine rifampicin resistence of *Mycobacterium tuberculosis* directly from sputum. Int. J. Tuberc. Lung Dis.8: 1114-1119.

Albert, H., Heyderyde, A., Mole, R., Trollip, A., and Blumberg, L. (2001). Evaluation of fast plaque TB-RIF TM, a rapid, manual test for the determination of rifampicina resistance from *Mycobacterium tuberculosis* cultures. *Int J Tuberc Lung Dis;* 5 (10): 906-11

Ando H, Mitarai S, Kondo Y, Suetake T, Kato S, Mori T, and Kirikae T. (2011). Evaluation of a line probe assay for the rapid detection of *gyrA* mutations associated with fluoroquinolone resistance in multidrug-resistant *Mycobacterium tuberculosis*. Journal of Medical Microbiology. 60:184-88.

Angeby, K.A., Klintz, L., and Hoffner, S.E.. (2002). Rapid and inexpensive drug susceptibility testing of *Mycobacterium tuberculosis* with a nitrate reductase assay. *J Clin Microbiol;* 40: 553-5.

Arnvig Kristine B, Gopal B, Papavinasasundaram K. G, Cox Robert A., Colston M. Joseph. (2005). The mechanism of upstream activation in the *rrnB* operon of *Mycobacterium smegmatis* is different from the *Escherichia coli* paradigm. Microbiology.151, 467–473. http://mic.sgmjournals.org/content/151/2/467.full.pdf+html

Ashworth M, Horan K, Freeman R, Oren E, Narita M, and Cangelosi G. 2008. Use of PCR-Based *Mycobacterium tuberculosis* Genotyping To Prioritize Tuberculosis Outbreak Control Activities. Journal of Clinical Microbiology.46(3):856-62.

Bardarov S, Jr., Dou H, Eisenach K,Banaiee N, Ya S, Chan J. et al. (2003).Detection and drug-susceptibility testing of *Mycobacetrium tuberculosis* from sputum samples using Luciferase reporter phage: comparison with the mycobacteria Growth Indicator Tube (MGIT) system. Diagn Microbiol Infect Dis 45:53-61.

Bauer J, Andersen AB, Kremer K, et al. (1999). Usefulness of spolygotyping to discriminate IS6110 low-copy-number *Mycobacterium tuberculosis* complex strains cultured in Denmark. J Clin Microbiol.37:2602-2606.

Belén I.; Morcillo N. ; Bernardelli A. (2007). Identificación fenotípica de micobacterias. Bioquímica y Patología Clínica. 71(2):47-51.

Bentwich, Z., G. Maartens, D. Torten, A.A. Lal, and R.B. Lal. (2000). Concurrent infections and HIV pathogenesis. AIDS.14:2071-81.

Bernardelli Amelia. (2007). Manual de Procedimientos. Clasificación fenotípica de las micobacterias. Dirección de Laboratorio y Control Técnico. Available on:http://www.senasa.gov.ar/Archivos/File/File1443- mlab.pdf-BioSource International, Inc. Alamar blue™ ordening information. Catalog number DAL 1100.

Blaschitz M, Hasanacevic D, Hufnagil P, Hasenber P, Pecavar V, meidlinger L, Konr M, Allerberger F and Infra A. (2011). Real-time PCR for single-nucleotide polymorphism detection in the 16S rRNA gene as an indicator for extensive drug resistance in *Mycobacterium tuberculosis*. J Antimicrob Chemother. 66(6):1243-46.

Breen, R. A., et al.(2007). Rapid diagnosis of smear-negative tuberculosis using immunology and microbiology with induced sputum in HIV-infected and uninfected individuals PLoS One 2:e1135.

Brosch R, Gordon SV, Marmiesse M, et al. (2002). A new evolutionary scenario for the *Mycobacterium tuberculosis* complex. Proc Natl Acad Sci U S A. 99:3684-3689.

Burgos MV, Mendez JC, Ribon W. (2004). Molecular epidemiology of tuberculosis: methodology and applications. Biomédica. 24(supl.1):188-201

Buszewski, B., M. Kesy, T. Ligor, and A. Amann. (2007). Human exhaled air analytics: biomarkers of diseases. Biomed. Chromatogr. 21:553-566.

Butt, T., R. N. Ahmad, R. K. Afzal, A. Mahmood, and M. Anwar. (2004). Rapid detection of rifampicin susceptibility of *Mycobacterium tuberculosis* in sputum specimens by mycobacteriophage assay. J pak. Med. Assoc. 54:379-382.

Caminero JA, Pena MJ, Campos-Herrero MI, et al. (2001). Exogenous reinfection with tuberculosis on a European island with a moderate incidence of disease. Am J Respir Crit Care Med.163:717-720.

Caminero J.A. (2003). Guía de la tuberculosis para médicos especialistas. París: Unión Internacional Contra la Tuberculosis y Enfermedades Respiratorias (UICTER).

Canettii, G., S. Froman, J. Grosset, P. Hauduroy, M. Langerova, H. T. Mahler, G. Meissner, D. A. Mitchison, and L. Sula. (1963). Mycobacteria: laboratory methods for testing drug sensitivity and resistance. *Bull. W. H. O.*;29:565–78

Canettii, G., W. Fox, A. Khomenko, H. T. Mahler, N. K. Menon, D. A. Mitchison, N. Rist, and N. A. Smelev. (1969). Advances in techniques of testing mycobacterial drug sensitivity and the use of sensitivity tests in tuberculosis control programs. *Bull. W. H. O.*; 41:21–43.

Cardoso S., Martin A., Mejia G., Palomino J., Da Silva M., Portaels F. (2004). Practical handbook for the phenotypic and genotypic identification of mycobacteria. Section 1.

Carriere C, Riska PF, zimhony O, Kriakov J, Bardarov S, Burns J, et al. (1997). Conditionally replicating luciferase reporter phages: improved sensitivity for rapid detection and assessment of drug susceptibility of *Mycobacterium tuberculosis*. J Clin Microbiol. 35:3232-9.

Castiblanco C.A., and W. Ribón. (2006). Coinfección de tuberculosis en pacientes con VIH/SIDA: un análisis según las fuentes de información en Colombia. *Infectio*; 10(4): 232-42.

Castro Claudia, Soler-Tovar Diego, Brieva-Rico Claudia, Moreno Martha, Ribón Wellman. (2010). Micobacteriosis aviar en un búho orejudo *Asio stygius*, de origen urbano en Bogotá, Colombia. In: Aves rapaces y conservación. Una perspectiva Iberoameriana. Victor Hernandez, Ruth Muñiz, José Cabot y Tjitte de Vries.43-48. Tundra Ediciones. ISBN 978-84-937873-2-5. Valencia, España.

Castro CM, Puerto G, García LM, Orjuela DL, Llerena C, Garzón MC, Ribón W. (2007). Identificación molecular de micobacterias no tuberculosas mediante el análisis de los patrones de restricción, Colombia 1995-2005. Biomédica. 27:439-46

Caviedes, L., Lee, T.S., and Gilman, R.H. (2000). Rapid, efficient detection and drug susceptibility testing of *Mycobacterium tuberculosis* in sputum by microscopic observation of broth cultures. The Tuberculosis Working Group in Peru. *J Clin Microbiol*; 38: 1203-8.

Centers for Disease Control and Prevention (CDC). (2006). Emergence of *Mycobacterium tuberculosis* with Extensive Resistance to Second-Line Drugs - Worldwide, 2000-2004. MMWR; 55:301-8.

Centers for Disease Control and Prevention (CDC). (2007). Extensively drug-resistant tuberculosis – United States, 1993- 2006. *MMWR*; 56:250-3.

Clinical and Laboratory standards Institute. (2003).Suceptibility testing of Mycobacteria, Nocardiae, and other aerobic Actinomycetes; approved standard. CLSI document M24-A. Clinical and Laboratory Standars Institute,Wayne, PA.

Cockerill, F. R., III. (1999). Genetic methods for assessing antimicrobial resistance. Antimicrob. Antimicrob. Agents Chemother. 43:199-212.

Coll, P. (2003). Fármacos con actividad frente a *Mycobacterium tuberculosis*. *Enferm Infecc Microbiol Clin*; 21:299-308.

Collins D. (2011). Advances in molecular diagnostics for *Mycobacterium bovis*. Veterinary Microbiology. 151:2-7

Coros A, DeConno E, Derbyshire KM.(2008). IS6110, a *Mycobacterium tuberculosis* complex-specific insertion sequence, is also present in the genome of *Mycobacterium smegmatis*, suggestive of lateral gene transfer among mycobacterial species. J Bacteriol. 190:3408-3410.

Daley, P., et al. (2009). Blinded evaluation of comercial urinary Lipoarabinomannan for active tuberculosis: a pilot study. Int. J. Tuberc. Lung Dis. 13:989-995

De Beenhouwer, H., et al. (1995). Rapid detection of rifampicin resistance in sputum and biopsy specimens from tuberculosis patients by PCR and line probe assay. Tuber. Lung Dis. 76:425-430

De Rossi E, Ainsa JA, Riccardi G. (2006). Role of mycobacterial efflux transporters in drug resistance: an unresolved question. *FEMS Microbiol Rev*; 30: 36-52.

De Viedman, D. G. (2003). Rapid detection of resistance in *Mycobacterium tuberculosis*: a review discussing molecular approaches. Clin. Microbiol. Infect. 9:349-359.

Deng S. Yuan T, Xia J, Huang H, Cheng X, Chen M.(2011). Clinical utility of a combination of lipoarabinomannan, 38-kDa, and 16-kDa antigents as a diagnosis tool for tuberculosis. Diagnostic Microbiology and Infectious Disease. 71:46-50.

Diaz R. (2003). "Caracterización molecular de cepas de *Mycobacterium tuberculosis* y su implicación en el control de la tuberculosis en Cuba. Habana: Instituto De Medicina Tropical "Pedro Kourí. 2-83.

Diel R, Schneider S, Meywald-Walter K, et al. (2002). Epidemiology of tuberculosis in Hamburg, Germany: long-term population-based analysis applying classical and molecular epidemiological techniques. J Clin Microbiol. 40:532-539.

Djiba, M.L., A.I. Sow, M. Ndiaye, and J.A. Dromigny. (2004). Evaluation of E-test method in determination of susceptibility of *Mycobacterium tuberculosis* to four antituberculosis drugs. *Dakar Med*; 49(3):185-91.

Doroudchi M, Kremer K, Basiri EA, et al. (2000). IS6110-RFLP and spoligotyping of *Mycobacterium tuberculosis* isolates in Iran. Scand J Infect Dis. 32:663-668.

Ecker, D. J., et al. (2008). Ibis T5000 a universal biosensor approach for microbiology. Nat. Rev. Microbiol. 6:553-558.

Ecker, D. J., et al. (2009). Molecular genotyping of microbes by multilocus PCR and mass spectrometry: a new tool for hospital infection control and public health surveillance. Methods Mol. Biol. 551:71-87.

Eltringham IJ, Drobniewski FA, Mangan JA, Butcher PD, Willson SM. (1999).Evaluation of reverse transcription-PCR and bacteriophage-based assay for rapid phenotypic detection of rifampin resistance in clinical isolates of *Mycobacterium tuberculosis*. J Clin Microbiol. 37:3528-32.

Eltringham IJ, Wilson SM, DrobniewsKi FA. (1999). Evaluation of a bacteriophage-based assay (phage amplified biologically assay) as a rapid screen for resistance to isoniazid, ethambutol streptomycin, pyrazinamide, and ciprofloxacin among clinical isolates of *Mycobacterium tuberculosis*. J Clain Microbiol. 37:3528-32.

Eshoo, M. W., et al. (2010). Detection and identification of Escherichia species in blood by use of PCR and electrospray ionization mass spectrometry. J. Clin. Microbiol. 48:472-478

Evans JT, Hawkey PM, Smith EG, et al. (2004). Automated high-throughput mycobacterial interspersed repetitive unit typing of *Mycobacterium tuberculosis* strains by a combination of PCR and nondenaturing high-performance liquid chromatography. J Clin Microbiol. 42:4175-4180.

Fan XY, Hu ZY, Xu FH,Yan ZQ, Guo SQ, Li ZM. (2003).Rapid detection of *rpo*B gene mutations in rifampin-resistant *Mycobacterium tuberculosis* isolates in Shanghai by using the amplification refractory mutation system. J Clin Microbiol. 41:993-7.

Fluit, A. C., M. R. Visser, and F. J. Schmitz. (2011). Molecular detection of antimicrobial resistence. Clin. Microbiol. Rev. 14:836-871.

Fisher Mark.(2002).Diagnosis of MDR-TB: a developing world problem on a developed world budget. Expert Review of Molecular Diagnostics.2(2):151-159. Available on: http://www.expert- reviews.com/doi/abs/10.1586/14737159.2.2.151

Gandhi, N.R., A. Moll, and A.W. Sturm. (2006). Extensively drug-resistant tuberculosis as a cause of death in patients co- infected with tuberculosis and HIV in a rural area of South Africa. *Lancet*; 368: 1575-80

Garcia de Viedman D, del Sol Diaz Infantes M, Lasala F, Chaves F, Alcala L, Bouza E. (2002).New real-time PCR able to detect in a single tube multiple rifampin resistance mutations and high-level isoniazid resistence mutations in *Mycobacterium tuberculosis*. J Clin Microbiol. 40:988-95.

Gegia, M., et al. (2008). Prevalence of and molecular basis for tuberculosis drug resistance in the Republic of Georgia: validation of a QIAplex system for detection of drug resistance-related mutations. Antimicrob. Agents Chemother. 52:725-729.

Gibson A, Brown T, Baker L, et al. (2005).Can 15-locus mycobacterial interspersed repetitive unit-variable-number tandem repeat analysis provide insight into the evolution of *Mycobacterium tuberculosis*? Appl Environ Microbiol. 71:8207-8213.

Gibson AL, Huard RC, Gey van Pittius NC, et al. (2008).Application of sensitive and specific molecular methods to uncover global dissemination of the major RDRio Sublineage of the Latin American-Mediterranean *Mycobacterium tuberculosis* spoligotype family. J Clin Microbiol. 46:1259-1267.

Gillespie SH. (2002). Evolution of drug resistance in *Mycobacterium tuberculosis:* clinical and molecular perspective. *Antimicrob Agents Chemother*;46:267-74.

Gopaul KK, Brown TJ, Gibson AL, et al. (2006).Progression toward an improved DNA amplification-based typing technique in the study of *Mycobacterium tuberculosis* epidemiology. J Clin Microbiol. 44:2492-2498.

Gordon SV, Heym B, Parkhill J, et al. (1999).New insertion sequences and a novel repeated sequence in the genome of *Mycobacterium tuberculosis* H37Rv. Microbiology.145 (Pt 4):881-892.

Grant, R. J., et al. (2010). Application of the Ibis-T5000 pan-orthopoxvirus assay to quantitatively detect monkeypox viral loads in clinical specimens from macaques experimentally infected with aerosolized monkeypox virus. Am. J. Trop. Med. Hyg. 82:318-323

Hazbon M. (2004). Recent advances in molecular methods for early diagnosis of tuberculosis and drug-resistant tuberculosis. Biomédica.24:149-62.

Heifets, L. (2000). Conventional methods for antimicrobial susceptibility testing of *Mycobacterium tuberculosis*. In: Multidrug- resistant Tuberculosis, Ed.: Bastian I, Portaels F. Kluwer Academic Publishers, Dordrecht, The Netherlands.

Heifets, L., and Sanchez, T. (2003). Agar medium for the growth of *Mycobacterium tuberculosis*. United States Patent 6579694.

http://www.freepatentsonline.com/6579694.html

Heifets, L.B., and G.A. Cangelosi. (1999). Drug susceptibility testing of *Mycobacterium tuberculosis:* a neglected problem at the turn of the century. *Int J Tuberc Lung Dis*; 3: 564-81.

Hernández R, Agilar D, Cohen I, Guerrero M, Ribón W, Acosta P, Orozco H, Marquina B, Salinas C, Rembao D, Espitia C. (2010). Specific bacterial genotypes of *Mycobacterium tuberculosis* cause extensive dissemination and brain infection in an experimental model. Tuberculosis. 90 268- 277.

Ho, Y. P., and P. M. Reddy.(2010). Indentification of pathogens by mass spectrometry. Clin. Chem. 56:526-536. (Epub ahead of print.)

Janossy, G. (2008). The changing pattern of "smart" flow cytometry (S-FC) to assist the cost-effective diagnosis of HIV, tuberculosis, and leukemias in resource-restricted conditions. Biotechnol. J. 3:32-42.

Janossy, G., et al. (2008). The role of flow cytometry in the interferongamma-based diagnosis of active tuberculosis and its coinfection with HIV-1 — a technically oriented review. Cytometry B Clin. Cytom. 74:S141-S151.

Jones WD, Jr. (1975).Differentiation of known strains of BCG from isolates of *Mycobacterium bovis* and *Mycobacterium tuberculosis* by using mycobacteriophage 33D. J Clin Microbiol.1:391-392.

Jones WD, Jr., Greenberg J. (1978). Modification of methods used in bacteriophage typing of *Mycobacterium tuberculosis* isolates. J Clin Microbiol. 7:467-469.

Kalantri S, Pai M, Pascopella L, et al. (2005).Bacteriophage- based tests for the detection of *Mycobacterium tuberculosis* in clinical specimens: a systematic review and meta-analysis. BMC Infect Dis. 5:59.

Kamerbeek J, Schouls L, Kolk A, et al. (1997).Simultaneous detection and strain differentiation of *Mycobacterium tuberculosis* for diagnosis and epidemiology. J Clin Microbiol. 35:907-914.

Kassama Y, Shemko M, Shetty N, et al. (2006). An improved fluorescent amplified fragment length polymorphism method for typing *Mycobacterium tuberculosis*. J Clin Microbiol. 44:288-289.

Kent, P. T.,and G. P. Kubica. (1985). Public Health mycobacteriology: a guide for the level III Laboratory. Centers for Disease Control, Atlanta, GA.

Kim BJ, Kim SY, Park BH, Lyu MA, Park IK, Bai GH, et al. (1997).Mutations in the *rpo*B gene of *Mycobacterium tuberculosis* that interfere with PCR-single-strand conformation polymorphism analysis for rifampin susceptibility testing. J Clin Microbiol.35:492-4.

Kirschner P, Springer U, Vogel A, Meier A, Wrede M, Kiekenbeck F, Bange and Bottger E. (1993). Genotypic identification of mycobacteria by nucleic acid sequence determination: report of a 2-year experience in a clinical laboratory. J Clin Microbiol 31:2882-9

Kluge, C., M. C. Gutierrez, K. Delabre, and G. Marchal. (2008). Improving detection of mycobacteria by immunocapture-qPCR in water samples: a rapid and sensitive method for routine testing, abstr. OP-24. 29th Annu. Congr. Eur. Soc. Micobacteriol., Plovdiv, Bulgaria.

Kremer K, Arnold C, Cataldi A, Gutierrez M. Haas W, Panaiotov S, Skuce R, Supply P, van der Zanden A. and van Soolingen D.. (2005).Discriminatory power and reproducibility of novel DNA typing methods for *Mycobacterium tuberculosis* complex strains. J Clin Microbiol. 43:5628-5638.

Lakshmi, V., Patil, M.A., Subhadha, K., and Himabindu, V. (2006). Isolation of mycobacteria by Bactec 460 TB system from clinical specimens. *Indian J Microbiol*; 24(2):124-6

Lawn, S. D., et al. (2009). Urine lipoarabinomannan assay for tuberculosis screening screening before antiretrobarial therapy: diagnostic yield and association with immune reconstitution disease. AIDS 23:1875-188.

Lee, H., T.J.Yoon, R weissleder. (2009).Ultrasentive detection of bacteria using corel-shell nano-particulas and an NMR – filter system. Ang.

Lemus, D., A. Martin, E. Montoro, F. Portaels, and J.C. Palomino. (2004). Rapid alternative methods for detection of rifampicin resistance in *Mycobacterium tuberculosis. J Antimicrob Chemoth*; 54: 130–3

Lemus, D., Montoro, E., Echemendía, M., Martin, A., Portaels, F., and Palomino, J.C. (2006). Nitrate reductase assay for detection of drug resistance in *Mycobacterium tuberculosis*: simple and inexpensive method for low-resource laboratories. *J Med Microbiol*; 55: 861–3

Li D, Wang T, Song X, Qucuo M, Yang B, Zhang J, Wang J, Ying B, Tao C, Wang L. (2011). Genetic study of two single nucleotide polymorphisms within corresponding microRNAs and susceptibility to tuberculosis in a Chinese Tibetan and Han population. Human Immunology. 72:598-602.

Lin, S. Y., E. Desmond, D. Bonato, W. Gross, and S. siddiqi. (2009). Multicenter evaluation of Bactec MGIT 960 system for second-line drug susceptibility testing of *Mycobacterium tuberculosis* complex. J. Clin. Microbial. 47:3630-3634

LishansKi A, Kurn N, Ullman EF. (2000).Branch migration inhibition in PCR-amplified DNA: homogeneous mutation detection. Nucleic Acids Res. 28:E42.

LishansKi A. (2000). Screening for single-nucleotide polymorphism using branch migration inhibition in PCR-amplified DNA. Clin Chem. 28:E42.

Lyiu YP, Behr MA, Small PM, Kurn N. (2000). Genotypic determination of mycobacterium tuberculosis antibiotic resistance using a novel mutation detection method, the branch migration inhibition *Mycobacterium tuberculosis* antibiotic resistance test. J Clin Microbiol. 38:3656-62.

Long, R. (2000). Drug-resistant tuberculosis. CMAJ;163(4):425-8.

Maeda H., S. Matsu-Ura, Y. Yamauchi, and H. Ohmori. (2001). Resazurin as an electrón aceptor in glucosa oxidasecatalized oxidation of glucose. *Chem. Pharm. Bull.*; 49(5):622-5.

Martin, A., M. Camacho, F. Portaels, and J.C. Palomino. (2003). Resazurin Microtiter Assay Plate Testing of *Mycobacterium tuberculosis* susceptibilities to Second-Line Drugs: Rapid, Simple, and Inexpensive Method. *Antimicrob Agents Chemother*; 47(11): 3616–9

Mejía, G.I., Guzmán, A., Agudelo, C.A., Trujillo, H., and Robledo, J. (2004). Cinco años de experiencia con el agar de capadelgada para el diagnóstico rápido de tuberculosis. *Biomédica*; 24(1)

Mitchison, D.A. (2005). Drug resistance in tuberculosis. *Eur Respir J*; 25: 376–9

Mokrousov I, NarvsKaya O, Limeschenko E, Otten T, VyshnevsKiy B. (2002). Detection of ethambutol-resistant *Mycobacterium tuberculosis* strains by multiplex allele-specific PCR assay targeting embB306 mutations. J Clin microbial. 40:1617-20.

Mokrousov I, Otten T, FilipenKo M, Vyazovaya A, Chrapov E, Limescheko e, et al. (2002).detection of isoniazid-resistant *Mycobacterium tuberculosis* strains by a multiplex allele specific PCR assay targeting KatG codon 315 variation. J Clin Microbiol. 40:2509-12.

Mokrousov I, Otten T, VyshnevsKiy B, NarvsKaya O. (2002).Detection of embB306 mutations in ethambutol-susceptible clinical isolates of *Mycobacterium tuberculosis* from Northwestern Russia: implications for genotypic resistance testing. J Clin microbial.40:3810-3.

Montoro, E., Lemus, D., Echemendia, M., Martin, A., Portaels, F., and Palomino, J.C. (2005). Comparative evaluation of the nitrate reduction assay, the MTT test, and the resazurin microtitre assay for drug susceptibility testing of clinical isolates of *Mycobacterium tuberculosis*. *Antimicrob Agents Chemother*; 55: 500–5

Morcillo, N., B.Di Giulio, B. Testani, M. Pontino, C. Chirico, and A. Dolmann. (2004). A Microplate Indicator-based method for determining Drug-Susceptibility of Multidrug-Resistant *Mycobacterium tuberculosis* to antimicrobial agents. *Int J Tuberc Lung Dis.* 8() 253-259.

Mortimer P, Arnold C. (2001). FAFLP: last word in microbial genotyping? J Med Microbiol. 50:393-5.

Mossmann T. (1983). Rapid colorimetric assay for cellular growth and survival: application to proliferation and cytotoxicity assays. *J Immunol Methods*; 65:55-63.

Mostrom P, Gordon M, Sola C, Ridell M, Rastogui N.(2002). Methods used in the molecular epidemiology of tuberculosis. Clin Microbiol Infect. 8:694-704.

Musser, J.M. (1995). Antimicrobial agent resistance in mycobacteria: molecular genetic insights. *Clin. Microbiol. Rev.* 8:496– 514.

Mutetwa, R., et al. (2009). Diagnostic accuracy of commercial urinary lipoarabinomannan detection in African tuberculosis suspects and patients. Int. J. Tuberc. Lung Dis. 13:1253-1259.

Narayanan S. (2004). Molecular epidemiology of tuberculosis. Indian J Med Res. 120:233-247.
National TB Controllers Association / CDC Advisory Group on Tuberculosis Genotyping. (2004). Guide to the Application of Genotyping to Tuberculosis Prevention and Control. Atlanta: Department of Health and Human Services, CDC.
Niobe-Eyangoh SN, Kuaban C, Sorlin P, et al. (2005). Molecular characteristics of strains of the Cameroon family, the major group of *Mycobacterium tuberculosis* in a country with a high prevalence of tuberculosis. J Clin Microbiol. 42:5029-5035.
O'Brien J., Wilson I., Otor T., and Pognan F. (2000). Investigation of the alamar blue (resazurin) fluorescent dye for the assessment of mammalian cell cytotoxicity. *Eur J Biochem*; 267:5421-6.
Organización Panamericana de la Salud. (2008). Normas y Guía Técnica. Manual para el Diagnóstico Bacteriológico de la Tuberculosis. Capítulo II. Cultivo.
Orjuela D, Puerto G, Mejía G, Castro C, Garzón M C,, García L M, Hernández E, Ribón W, Rodríguez G. (2010). Cutaneous tuberculosis after mesotherapy: report of six cases. Biomédica. 30:321-6.
Pai, M., and Kalantri, S.P. (2005). Bacteriophage-Based tests for tuberculosis. *Indian J Med Microbiol*; 23(3):149-50
Pai, M., S. Kalantri, L. pascopella, L. W. Riley, and A. L. Reingold. (2005). Bacteriophage-based assays for the rapid detection of rifampicin resistance in *Mycobacterium tuberculosis*: a meta-analisys. J. Infect. Dis. 51:175-187.
Palomino, J.C. (2005). Non conventional and new methods in the diagnosis of tuberculosis: feasibility and applicability in the field. *Eur Respir J*; 26:1-12.
Palomino, J.C., A. Martin, M. Camacho, H. Guerra, J. Swings, and F. Portaels. (2002). Resazurin Microtiter Assay Plate: Simple and Inexpensive Method for Detection of Drug Resistance in *Mycobacterium tuberculosis*. *Antimicrob Agents Chemother*; 46(8a): 2720–2
Palomino, J.C., and Portaels, F. (1999). Simple procedure for drug susceptibility testing of *Mycobacterium tuberculosis* using a commercial colorimetic assay. *Eur J Clin Microbiol Infect Dis*; 18: 380-3.
Palomino, J.C., S.C Leão, and V. Ritacco. (2007). Tuberculosis 2007. First Edition. www.Tuberculosistextbook.com.
Parra A. (2003).Epidemiología de la Tuberculosis en Artiodáctilos Salvajes de Extremadura. Universidad De Extremadura.
Parson, L. M., A. Somoskovi, R. Urbanczik, and M. Salfinger. (2004).Laboratory diagnostic aspects of drug resistant tuberculosis. Front. Biosci. 9:2086-2105
Parsons LM, Somoskövi A, Urbanczik R, Salfinger M. (2004). Laboratory diagnostic aspects of drug resistant tuberculosis. *Front Biosci*;1:2086-105.
Pavlou, A. K., and A. P. Turner. (2000). Sniffing out the truth: clinical diagnosis using the electronic nose. Clin. Chem. Lab. Med. 38:99-112.
Pavlou, A. K.,et al. (2004). Detection of *Mycobacterium tuberculosis* (TB) in vitro and in situ using and electronic nose in combination with a neural network system. Biosens. Bioelectron. 20:538-544.
Pfyffer, G. (2007). *Mycobacterium*: general characteristics, laboratory detection, and staining procedures, p. 543-572. In P. R. Murray, E. J. Baron, J. H. Jorgensen, M. L. Landry, and M. A. Pfaller (ed.), Manual of clinical Microbiology, 9th ed. ASM Press, Washington, DC.

Pfyffer, G.E., D.A. Bonato, and A. Ebrahimzadeh. (1999). Multicenter laboratory validation of susceptibility testing of *Mycobacterium tuberculosis* against classical second-line and newer antimicrobial drugs by using the radiometric BACTEC 460 technique and the proportion method with solid media. *J Clin Microbiol*; 37: 3179-86

Pfyffer, G.E., H.M. Welscher, and P. Kissling. (1997). Comparison of the Mycobacteria Growth Indicator Tube (MGIT) with radiometric and solid culture for recovery of acid-fast bacilli. *J Clin Microbiol*; 35: 364-8.

Phillips, M., et al. (2007). Volatile biomarkers of pulmonary tuberculosis in the breath. Tuberculosis 87:44-52.

Piatek, A. S., et al (2000). Genotypic analisis *Mycobacterium tuberculosis* in two distinct populations using molecular beacons: implications for rapid susceptibility testing. Antimicrob. Agents chemother. 44:103-110.

Piatek, A. S., et al. (1998). Molecular beacon sequence analysis for detecting drug resistance in *Mycobacterium tuberculosis*. Nat. Biotechnol. 16:359-363.

Pino, P.P., C. Gassiot, J.C. Rodríguez, I. Páez, J. Barreto, and J. Gundián. (1998). Tratamiento de la tuberculosis resistente a múltiples drogas. *Acta médica*;8(1):110-7.

Prodhom G, Guihot C, Gutierrez M, Varnerot A, Gicquel B, Vincent V. (1997). Rapid discrimination of *Mycobacterium tuberculosis* complex strains by lugation-mediated PCR fingerprint analysis. J Clin Microbiol. 35:3331-4.

Puerto G. Castro CM, Ribón W. (2007).Reacción en cadena de la polimerasa: una contribución para el diagnóstico de la tuberculosis extrapulmonar y de las micobacteriosis. Infectio.11(2):87-94.

Raizada, N., et al. (2009). Establishing laboratory proficiency to conduct line probe assay For MDR-TB diagnosis. Int. J. Tuberc. Lung DIs. 13(s1):309.

Ramaswamy S, Musser JM. (1998).Molecular genetic basis of antimicrobial agent resistance in *Mycobacterium tuberculosis*: 1998 update. Tuber Lung Dis.79:3-29.

Raviglione, M., and I. Smith. (2007). XDR Tuberculosis---Implications for Global Public Health. *N. Engl. j. med*; 356(7): 656- 9.

Reisig F, Kremer K, Amthor B, van Soolingen D, and Haas W.(2005). Fast Ligation Mediated PCR a Fast and Reliable Method for IS6110 Base Typing of *Mycobacterium tuberculosis* Complex. Journal of Clinical Microbiology. 43(11):5622- 27.

Reither, K., et al. (2009). Low sensitivity of a urine Lam-ELISA in the diagnosis of pulmonary tuberculosis. BMC Infect. Dis. 9:141.

Richardson, M., et al. (2002). Multiple *Mycobacterium tuberculosis* strains in early cultures from patients in a high-incidence community setting. J. Clin. Microbiol. 40:2750-2754.

Riska PF, Jacobs WR, Jr. (1998).The use of luciferse-reporter phage for antibiotic-susceptibility testing of mycobacteria. Methods Mol Biol.101:431-55.

RisKa PF, Jacobs WR, Jr., Bloom BR, McKitricK J, Chan J. (1997).Specific identification of *Mycobacterium tuberculosis* with the luciferase reporter mycobacteriophage: use of P-Nitro-alpha-acetylamino-beta-hydroxy propiophenone. J Clin Microbiol.35:3225-31.

RisKa PF, Su Y, Bardarov S, Freundlich L , SarKis G, Hatfull G, et al. 1999.Rapid film-based determination of antibiotic susceptibilities of *Mycobacterium tuberculosis* strains by using a luciferase reporter phage and the Bronx box. J Clin Microbiol.37:1144-9.

Rivoire, N., P. Ravololonandriana, T. Rasolonavalona, A. Martin, F. Portaels, H. Ramarokoto, and V.R. Razanamparany. (2007). Evaluation of the resazurin assay for the detection of multidrug- resistant *Mycobacterium tuberculosis* in Madagascar. *Int J Tuberc Lung Dis*; 11(6):663-88.

Robledo, J., G. I. Mejia, L. Paniagua, A. Martin, and A. Guzman. 2008. Rapid detection of rifampicin and isoniazid resistance in *Mycobacterium tuberculosis* by the direct thin-layer agar method. Int. J. Tuberc. Lung Dis. 12:1482-1484.

Roth A, Fisher M, Hamid E, Michalke S, Ludwig W and Mauch H, (1998). Differentiation of phylogenetically related slowly growing mycobacteria based on 16S-23S rRNA gene internal transcribed spacer sequences. J Clin Microbiol 36:139-47.

Rozo-Anaya Juan C, Ribón Wellman. (2010). Molecular tools for *Mycobacterium tuberculosis* genotyping. Rev. salud pública. 12 (3): 510-521.

Shamputa IC, Van Deun A, Salim AH, et al. (2007).Endogenous reactivation and true treatment failure as causes of recurrent tuberculosis in a high incidence setting with a low HIV infection. Tropical Medicine & International Health.12:700-708.

Siddiqi, S. H.,and S. Rusch-Gerdes. (2006). MGIT procedure manual. Foundation for Innovative New Diagnostics, Geneva, switzarland.

Siddiqi, S.H., Libonati, J.P., and Middlebrook G.. (1981). Evaluation of rapid radiometric method for drug susceptibility testing of *Mycobacterium tuberculosis*. *J. Clin. Microbiol*; 13:908–12.

Sims E. Goyal M. Arnold C. (2002). Experimental versus in silico fluorescent amplified fragment length polymorphism analysis of *Mycobacterium tuberculosis* improved typing with an extended fragment range. J Clin Microbiol. 40:4072- 6.

Small PM, Hopewell PC, Singh SP, et al. (1994).The epidemiology of tuberculosis in San Francisco. A population-based study using conventional and molecular methods. N Engl J Med. 330:1703-1709.

Somoskovi, A., et al. (2003). Use of molecular methods to identify the *Mycobacterium tuberculosis* complex (MTBC) and other mycobacterial species and to detecd rifampin resistance in MTBC isolates following growth detection with the BACTEC MGIT 960 system. J. Clin. Microbiol. 41:2822-2826.

Steinlein L, Crawford j. (2001). Reverse Do Blot Assay (insertion Site Typing) for Precise Detection of Sites of IS6110 Insertion in the *Mycobacterium tuberculosis* genome. Journal of Clinical Microbiology. 39(3):871-78-

Streitz, M., et al (2007). Loss of receptor on tuberculin-reactive T-cells marks active pulmonary tuberculosis. PLoS One 2:e735.

Supply P, Lesjean S, Savine E, et al. (2001).Automated high-throughput genotyping for study of global epidemiology of *Mycobacterium tuberculosis* based on mycobacterial interspersed repetitive units. J Clin Microbiol. 39:3563-3571.

Supply P, Mazars E, Lesjean S, et al. (2000).Variable human minisatellite-like regions in the *Mycobacterium tuberculosis* genome. Mol Microbiol. 2000;36:762-771.

Tasso MP.; Martins MC.; Mizuka SY; Saraiva CM.; Silva MA. (2003). Cord formation and Colony morphology for the presumptive Identification of *Mycobacterium tuberculosis* complex. Brazilian Journal of Microbiology. 34:171-174.

Telenti A, Marchesi F, Balz M, Bally F, Bottger EC, Bodmer T. (1993).Rapid identification of mycobacteria to the species level by polymerase chain reaction and restriction enzyme analysis. J Clin Microbiol.31:175-8.

Telenti, A., P. Imboden, and F. Marchesi. (1993). Detection of rifampicin-resistance mutations in *Mycobacterium tuberculosis*. *Lancet*; 341: 647-50.

Telenti, A., P. Imboden, F. Marchesi, T. Schmidheini, and T. Bodmer. (1993). Direct, automated detection of rifampin resistant *Mycobacterium tuberculosis* by polymerase chain reaction and single-strand conformation polymorphism analysis. *Antimicrob Agents Chemother*; 37:2054-8.

Torres MJ, Criado A, Ruiz M, Llanos AC, Palomares JC, Aznar J. (2003).Improved real-time PCR for rapid detection of rifampin and isoniazid resistance in mycobacterium clinical isolates. Diagn Microbiol infect Dis.45:207-12.

Torres, M. J., A. criado, J. C. Palomares, and J. Aznar. (2000). Use of real-time PCR and fluorimetry for rapid detection of rifampin and isoniazid resistance-associated mutations in *Mycobacterium tuberculosis*. J. Clin. Microbiol. 38:3194-3199.

Torres-Duque C A, Díaz C, Vargas L, Serpa E M, Mosquera W, Garzón M C, Mejía G, García L M, González L A,. Castro C M, Ribón W. (2010).Disseminated mycobacteriosis affecting a prosthetic aortic valve: first case of *Mycobacterium peregrinum* type III reported in Colombia. Biomédica.30:332-7

Van SD, Hermans PW, de Haas PE, et al. (1991).Occurrence and stability of insertion sequences in *Mycobacterium tuberculosis* complex strains: evaluation of an insertion sequence-dependent DNA polymorphism as a tool in the epidemiology of tuberculosis. J Clin Microbiol. 29:2578-2586.

Varela G, Carvajales S, Gadea P, Sirok A, Grotiuz G, Mota M, Ritacco V, Reniero A, Rivas C, Schelotto F. (2005). Comparative molecular study of *Mycobacterium tuberculosis* strains, in times of antimicrobial drug resistance. Revista Argentina de Microbiología. 37:11-15.

Verma A., Avinash K., Wami JS.(1999). *Mycobacterium tuberculosis rrn* Promoters: Differential Usage and Growth Rate- Dependent Control. Journal of Bacteriology. 181(14):4326–4333.

Wayne, L.G. (1974). Simple pyrazinamidase and urease tests for routine identification of mycobacteria. *Am Rev Respir Dis*;109:147-51.

Wilson M. (2011). Recent Advances in the Laboratory Detection of *Mycobacterium tuberculosis* Complex and Drug. CID. 11:1350-55.

WoldeMeskel, D., G. Abate; M. Lakew, S. Goshu, A. Selassie, H. Miorner, and A. Aseffa. (2005). Evaluation of a direct colorimetric assay for rapid detection of rifampicin resistant *Mycobacterium tuberculosis*. *Ethiop.J.Health Dev*; 19(1):51- 4.

Wolfgan P. Gordon S. telenti A. and Cole S. (1998). Pulsed field Gel Electrophoresis for Mycobacteria. In: *Mycobacteria Protocols*. Tanya Paris and Neil G stoker. 51-63.Human Press.0-89603-471-2. Ttowa, New Jersey.

Woods, G., N. G. Warren, and C. B. Inderlied. (2007). Susceptibility test methods:Mycobacteria, Norcardia, and other Actynomicetes, p. 1223-1247. In P. R. Murray, E. J. baron, J. H. Jorgensen, M. L. Landry, and M. A. Pfaller (ed.),

World Health Organization. (2003). Treatment of tuberculosis: guidelines for national programmes. 3rd edition. Geneva.

World Health Organization. (2004). Anti-tuberculosis drug resistance in the world. Report N°3, prevalence and trends. Geneva. World Health Organization; WHO/HTM/TB/2004.343.

World Health Organization. (2006). Guidelines for the programmatic management of drug-resistant tuberculosis. Geneva, Switzerland: World Health Organization; WHO/HTM/TB/2006.361

World Health Organization. (2007). The Global MDR-TB y XDR-TB Response Plan 2007-2008. WHO/HTM/TB/2007.387.

World Health Organization. (2008). Interim policy guidance on drug susceptibility testing (DST) of second-line anti-tuberculosis drugs. WHO/HTM/TB/2008.392. World Health Organization, Geneva, Switzerland.

World Health Organization. (2009). Global tuberculosis control: Epidemiology, Estrategy and Financing. WHO Report 2009. Geneva, World Health Organization (WHO/HTM/TB/2009.411)

Yu M, chen H, Wu M, Huang L, Kuo Y, Yu F, and Jou R. (2011). Evaluation of the rapid MGIT TB. Identification Test for Culture Confirmation of *Mycobacterium tuberculosis* Complex Strain Detection.

Zhang Y, Mitchison D. (2003). The curious characteristics of pyrazinamide: a review. Int J Tuberc Lung Dis.7:6–21.

Zumia, A., and J. M. Grange. (2001). Multidrug-resistant tuberculosis — can the tide be turned? *Lancet Infect. Dis*; 1:199–202.

2

Rapid and Efficient Methods to Isolate, Type Strains and Determine Species of *Agrobacterium* spp. in Pure Culture and Complex Environments

Malek Shams, Tony Campillo, Céline Lavire,
Daniel Muller, Xavier Nesme and Ludovic Vial
Université de Lyon, Ecologie Microbienne Lyon,
UMR CNRS 5557, USC INRA 1193
France

1. Introduction

Agrobacterium are Alphaproteobacteria common in most soils that closely interact with plants in two respects. Firstly, and as a general trait of the whole taxon, they are rhizospheric bacteria saprophytically living in the root environment (i.e. rhizosphere) of numerous plants. Rhizospheric interactions are generally considered to be of commensal type with no detrimental effect to the plant, but in most instances they are likely beneficial to plants. For evident agronomic purposes it is worthwhile to explore whether agrobacteria are themselves plant growth-promoting rhizobacteria (PGPR) or not. However, this requires an expert determination of the *Agrobacterium* taxonomy. Indeed our current investigations suggest that only some agrobacterial species from the abundant soil *Agrobacterium* guild are selected in the rhizosphere of a given plant. Secondly, but only when they harbor a dispensable Ti plasmid (i.e. tumor inducing plasmid), agrobacteria are plant pathogens able to cause the crown gall disease to most dicots and gymnosperms and some monocots (Pitzscke & Hirt, 2010). Ti plasmids are conjugative and can easily spread in indigenous soil agrobacteria. As a result transconjugant agrobacteria become in turn pathogenic, contributing both to disease spread and perennial soil contamination. An epidemiological survey of crown gall thus also requires expert determination of the *Agrobacterium* taxonomy. A set of biochemical and molecular methods were thus set up to facilitate the taxonomic assessment of agrobacteria either in pure culture or directly from complex environments such as soils, rhizospheres or tumours. After a presentation of the present state of *Agrobacterium* taxonomy, this work provides efficient methods to : i) isolate agrobacteria from complex environments thanks to elective media; ii) determine the *Agrobacterium* genus status of newly isolated strains on the basis of a minimal set of biochemical tests; iii) determine species and type novel isolates of *Agrobacterium* by sequence analysis of relevant marker genes; iv) determine amplicon content and genome architecture; v) detect the presence of *Agrobacterium* and Ti or Ri plasmid directly in complex environments by PCR using selected primers and metagenomic DNA extracted from whole bacterial communities.

2. Current state of agrobacteria taxonomy

The *Agrobacterium* taxonomy was historically based upon pathogenicity traits that were later found to be determined by dispensable and highly exchangeable plasmids. It is now well known that crown gall and hairy root diseases — that are respectively characterized by tumors and root proliferations on infected plants — are due to the presence of infectious agrobacteria plasmids (i.e. Ti and Ri plasmids for tumor or root inducing plasmids, respectively) (Van Larebeke et al., 1974). However, agrobacteria were isolated and described very early — evidently before the discovery of the plasmidic nature of their pathogenicity. This is why first bacteria isolated from neoformed tissues, which had the ability to reproduce the symptoms, were named *Bacterium tumefaciens* and *Phytomonas rhizogenes*, respectively (Smith & Townsend, 1907; Riker, 1930). For their part, non pathogenic agrobacteria had been isolated earlier from soil and named *Bacillus radiobacter* because they displayed a particular star shape in certain growth conditions (Beijerinck & van Velden, 1902). The genus name *Agrobacterium* was created in 1942 by Conn as a genus very similar but different to the genus *Rhizobium* (Conn, 1942). Hooykaas et al. (1977) and Genetello et al. (1977) showed that Ti plasmid is transferable by conjugation in plasmid free agrobacteria that consecutively become able to incite crown gall. As a result, it was definitively evident that pathogenicity traits led to artificial classification of agrobacteria and could no longer be used to delineate species. In parallel, enzymatic abilities determined with biochemical galleries led to the distinction of several clusters among agrobacteria that were named biovars (Keane et al., 1970; Panagopoulos et al., 1978). Biovars are determined by chromosomal genes and not plasmids, thus tumorigenic, rhizogenic and non-pathogenic strains can be found within the same biovar.

In modern bacterial taxonomy, the criteria to delineate taxa now include genomic information. This has brought considerable modifications to the nomenclature. According to Wayne et al. (1987) and Stackebrandt et al. (2002), homogenous genomospecies is the ultimate criterion to validly delineate a *bona fide* species. Amongst agrobacteria *lato sensu*, genomospecies generally encompass the same set of strains than biovars. Upon this criterion, *Agrobacterium vitis* (i.e. biovar 3), *A. rubi* and *A. larrymoorei* have been validated as *bona fide* species (Ophel & Kerr, 1990; Holmes & Roberts, 1981; Bouzar & Jones, 2001). For its part biovar 1 is heterogeneous, including ten genomospecies (De Ley, 1973, 1974; Popoff et al., 1984) currently called genomovar G1 to G9 (Mougel et al., 2002) and G13 (Portier et al., 2006). Biovar 1 is thus a species complex and not a single bacterial species. Noticeably, G4 contains both type-strains of the formerly described species "*A. tumefaciens*" and "*A. radiobacter*" that should be both renamed *A. radiobacter* for antecedence reasons (Young et al., 2006). However, even though the name "*A. tumefaciens*" is not valid to designate a *bona fide* genomic species, we proposed that the group of closely related species corresponding to biovar 1 should be collectively called "*A. tumefaciens* complex" in order to avoid confusion with the genomic species G4, which must be validly named *A. radiobacter* (Costechareyre et al., 2010). For the ICSP subcommittee on the taxonomy of *Rhizobium* and *Agrobacterium*, this seems to be a good interim solution until genomovars can be formally named (Lindström & Young, 2011). In addition to the epithet *radiobacter* for genomovar G4, *fabrum* and *pusense* have recently been proposed for homogenous species corresponding to genomovar G8 and to genomovar G2, respectively (Lassalle et al., 2011; Panday et al., 2011).

Rapid and Efficient Methods to Isolate, Type Strains and Determine Species of Agrobacterium spp. in Pure Culture and Complex Environments

37

Modifications of the nomenclature also happen at the genus level. Although, biovar 2 is a homogenous genomospecies (Mougel et al., 2002; Popoff et al., 1984), this species clearly appears phylogenetically more related to *Rhizobium* species such as *R. etli* or *R. tropici*, than to other *Agrobacterium* species. Therefore, biovar 2 was proposed to be classified in the genus *Rhizobium*, while all other species were kept in the genus *Agrobacterium* (Costechareyre et al., 2010). This proposal had the great advantage to solve the paraphyletic problem of the old classification pointed out by Young et al. (2001). In agreement with taxonomy rule of name antecedence, biovar 2 should be named *Rhizobium rhizogenes*. The epithet for this species is *rhizogenes* because the type-strain ATCC 11322$5^T$ received this epithet at the time of its first description by Riker (1930). However, most members of this taxon are non-pathogenic or induce crown gall but not hairy root symptoms. This is particularly the case for the famous crown gall biocontrol agent K84 (Kerr, 1974). Frequently named "*A. radiobacter*" , it is actually a true member of the species *Rhizobium rhizogenes* in its present definition.

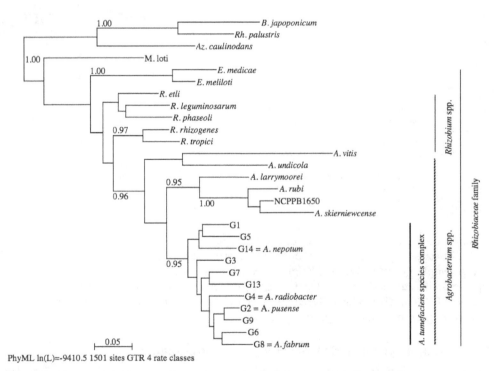

PhyML ln(L)=-9410.5 1501 sites GTR 4 rate classes

Fig. 1. Phylogeny of the *recA* gene of type-strains of all *bona fide* genomic species of *Agrobacterium* spp. known to date using the revised nomenclature proposed by Costechareyre et al. (2010). The maximum likelihood method was used. Only significant ML values (>0.95) are given. The branch length unit is the number of substitutions per nucleotidic site. *B. Bradyrhizobium, Rh. Rhodopseudomonas, Az. Azorhizobium, M. Mesorhizobium, E. Ensifer, R. Rhizobium, A. Agrobacterium.*

There are likely numerous other species of *Agrobacterium* either within or out the *A. tumefaciens* complex, such as the unnamed species that includes strain NCPPB 1650, or the novel species that got the epithet *skierniewicense* (Puławska et al., 2011) and—at least provisionally— *nepotum* for the novel *A. tumefaciens* genomovar G14 found by Puławska et al. (unpublished). In addition, it is also likely that bacteria able to induce the formation of nodules with some plants are true members of the genus *Agrobacterium* in its novel definition. It could be the case of the nitrogen fixing bacteria initially called *Allorhizobium undicola* (Costechareyre et al., 2010). A reasonably good view of the phylogeny of the *Agrobacterium* genus is given by the *recA* gene phylogeny (Figure 1) adapted from Costechareyre et al. (2010), completed with sequences from type-strains of novel species *A. skierniewicense* and *A. nepotum* (i.e. genomovar G14). The recent denomination *A. fabrum* and *A. pusense* are used for genomovars G8 and G2, respectively. Table 1 lists type-strains and *recA* accession numbers used to construct the phylogeny.

Genomic species	Type strain	LMG code	recA	Allele nb	Genome nb
Rhizobium rhizogenes	CFBP 2408[T]	150[T]	AM182126	nd	1
Agrobacterium vitis	K309[T]	8750[T]	AB253194	nd	1
A. undicola	ORS 992[T]	11875[T]	EF457952	nd	0
A. larrymoorei	AF3.10[T]	21410[T]	FN432355	4	0
A. rubi	TR3[T]	17935[T]	AM182122	2	0
A. sp.	NCPPB 1650[PT]	230	FN813466	1	0
A. skierniewcense	CH11[T]	2161[T]	HE610311	1	0
A. tumefaciens genomovar G1	TT111[PT]	196	FM164286	11	4
A. tumefaciens genomovar G2 = *A. pusense*	NRCPB10[T]	25623[T]	HQ166059	5	1
A. tumefaciens genomovar G3	CFBP 6623[PT]	nd	FM164304	2	1
A. tumefaciens genomovar G4 = *A. radiobacter*	ATCC 19358[T]	140[T]	FM164311	6	3
A. tumefaciens genomovar G5	CFBP 6626[PT]	nk	FM164318	3	2
A. tumefaciens genomovar G6	NCPPB 925[PT]	225	FM164319	1	1
A. tumefaciens genomovar G7	Zutra 3/1[PT]	198	FM164323	12	3
A. tumefaciens genomovar G8 = *A. fabrum*	C58[T]	287	FM164330	6	3
A. tumefaciens genomovar G9	Hayward 0363[PT]	27	FM164332	2	1
A. tumefaciens genomovar G13	CFBP 6927[PT]	nd	FM164333	1	1
A. tumefaciens genomovar G14 = *A. nepotum*	39/7[T]	26435[T]	(HE610312)	4	0

Table 1. Representative strains and infraspecific diversity of *Agrobacterium* species and related species. [T], type-strains. [PT], proposed type-strain. LMG, Laboratorium voor Microbiologie, Ghent University, Belgium. *recA*, accession number of *recA* allele of representative strain. (HE610312) is the *recA* accession number of strain C3.4.1. Allele nb, number of strains displaying different MLSA patterns. Genome nb, number of completely sequenced genome. nd, not deposited.

3. Isolation of agrobacteria from complex environments

Agrobacteria usually form ecological guilds in soils consisting of several genomovars with several strains within genomovars. There are also several strains even within a single tumor (Nesme et al., 1987; Vogel et al., 2003; Costechareyre et al., 2010). As a consequence, it is necessary to take extreme care during strain isolation to be sure of the genotypic purity of isolates. Efficient purification of agrobacteria requires both selective media and the strict respect of the procedure described below.

Rapid and Efficient Methods to Isolate, Type Strains and Determine Species of Agrobacterium spp. in Pure Culture and Complex Environments

39

Crown gall tumor on tomato plant   *Agrobacterium* colonies on 1A-Te medium

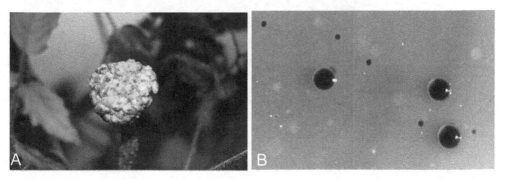

Fig. 2. Typical crown gall symptom (A) and *Agrobacterium* colony morphology on tellurite amended medium (B).

Elective media 1A and 2E, proposed by Brisbane and Kerr (1983) to isolate different biovars, are based on the particular ability of agrobacteria to use certain compounds, such as carbon and energy sources. In addition, these media contain selenite salt (Na_2SeO_3) because agrobacteria *lato sensu* has the general ability to reduce this salt. However, as the ability to reduce selenite also determines a general ability to resist to toxic amounts of tellurite (K_2TeO_3), this latter compound was substituted for selenite to improve the medium selectivity, especially when isolating bacteria from complex environments such as soil and decaying tumors (Mougel et al., 2001). Members of the *A. tumefaciens* species complex (i.e. biovar 1) and *A. rubi* can be isolated with 1A-Te containing arabitol as an elective carbon source and 80 mg/l K_2TeO_3. *R. Rhizogenes* (i.e. biovar 2) can be isolated using 2E-Te containing erythritol and 320 mg/l K_2TeO_3. Agrobacteria *lato sensu* (i.e. *Agrobacterium* spp. and *Rhizobium rhizogenes*) as well as most *Rhizobiaceae* can be isolated using MG-Te containing mannitol and 200 mg/l K_2TeO_3 (Table 2). On tellurite-amended media, colonies have the typical circular glistening morphologies of agrobacteria plus a characteristic black color with a metallic shine (Figure 2B).

Agrobacterium vitis (i.e. biovar 3) and *Agrobacterium larrymoorei* can be isolated on 3DG medium (Brisbane and Kerr, 1983) or Roy and Sasser medium (Roy and Sasser, 1983). On this latter medium, colonies have a typical pinkish-white (red centre) color. Both media are based on the utilization of sodium L-tartrate as the carbon source. It is likely that the selectivity of these media could be improved by addition of tellurite. This is because *A. vitis* members are resistant to this compound as well as most *Rhizobiaceae* (unpublished results).

For the first purification step, ca. 0.5 g of soil, root, shoot or tumor tissues are crushed with a micropestle in microtube containing 500 µl of sterile distilled water and then indispensably let to macerate for at least 30 min. This maceration is thought to be necessary to allow efficient cell separation from polysaccharides produced by agrobacteria. Next, ca. 25 µl of macerate (100 µl for soil samples) are streaked on appropriate selective media and then incubated at 28°C. Agrobacteria usually require between 3 and 4 days to form colonies on tellurite amended media because this compound tends to slow down bacterial growth. For the second purification step, clear individual colonies with the typical morphology (Figure 2B) are picked and suspended separately in 100 µl of sterile distilled water. These

1A-Te		2E-Te		MG-Te	
L-arabitol	3.04 g	Erythritol	3.05 g	D-Mannitol	5.0 g
NH₄NO₃	0.16 g	NH₄NO₃	0.16 g	L-glutamic acid	0.2 g
KH₂PO₄	0.54 g	KH₂PO₄	0.54 g	KH₂PO₄	0.5 g
K₂HPO₄	1.04 g	K₂HPO₄	1.04 g	NaCl	0.2 g
MgSO₄, 7H₂O	0.25 g	MgSO₄, 7H₂O	0.25 g	MgSO₄, 7H₂O	0.2 g
Sodium taurocholate	0.29 g	Sodium taurocholate	0.29 g	Yeast Extract	0.5 g
0.1% Crystal violet	2.0 ml	1 % Yeast Extract	1 ml	Agar	15 g
Agar	15 g	0.1 % Malachite green	5.0 ml	Adjust volume to 1 l with H₂O	
Adjust volume to 1 l with H₂O		Agar	15 g	Adjust to pH 7.2	
Sterilize by autoclaving		Adjust volume to 1 l with H₂O		Sterilize by autoclaving	
		Sterilize by autoclaving			
After autoclaving:		After autoclaving:		After autoclaving:	
K₂TeO₃	0.08 g	K₂TeO₃	0.32 g	K₂TeO₃	0.2 g
2 % cycloheximide	1 ml	2 % cycloheximide	1 ml	2 % cycloheximide	1 ml

Table 2. Improved selective media for agrobacteria members (adapted from Mougel et al., 2001).

individual colony suspensions are left to rehydrate overnight with shaking at 28°C before they are streaked onto appropriate media that must NOT contain tellurite. A third purification step following the same procedure is absolutely necessary before introduction of the novel isolate into the lab collection.

4. Presumptive genus determination by minimal biochemical tests

The INCO-DC European program ERBIC18CT970198, "Integrated Control of Crown Gall in Mediterranean Countries" has delivered a simple and efficient identification scheme for agrobacteria. By a simple and fast urease and/or esculinase biochemical test, any strains isolated from soil, tumor, or root that posseses the typical circulate glistening colony morphology on tellurite amended media can be conviently and confidently classified as part of the genus *Agrobacterium* (Table 3). To test the presence of these activities, fresh cells are suspended in urease solution or esculin solution (Table 3). After incubation at 28°C for 1 h, a colorimetric change is observed for agrobacteria strains: pink and black coloration, respectively, for urease and esculinase tests.

The *A. tumefaciens* species complex (biovar 1) has the enzymatic ability to aerobically convert lactose to 3-ketolactose. This is tested by streaking on medium containing lactose (Bernaerts and De Ley, 1963). After 2 days of growth at 28°C, the plates are flooded with a layer of Benedict's reagents. The presence of 3-ketolactose in the medium is indicated by the formation of a yellow ring around the growth of a positive strain (Table 3 and Figure 3).

A. fabrum (G8 genomovar of *A. tumefaciens* species complex), has the biovar 1 characteristics, plus a specific ability to degrade ferulic acid and caffeic acid. These properties are due to a gene cluster present only in the *A. fabrum* strains (Lassalle et al., 2011). A minimal medium supplemented with caffeic acid turns the color brown in five days. As caffeic acid degradation by *A. fabrum* inhibits this browning, inoculation of an isolate of agrobacteria in a minimal medium supplemented with caffeic acid is an easy test to detect if a strain belong to the *A. fabrum* species (Figure 3).

Rapid and Efficient Methods to Isolate, Type Strains and Determine Species of Agrobacterium spp. in Pure Culture
and Complex Environments

41

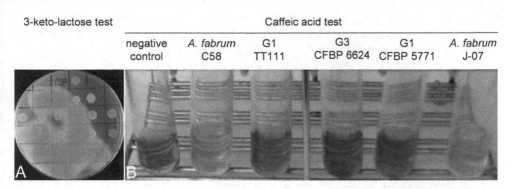

3-keto-lactose test	Caffeic acid test				
negative control	*A. fabrum* C58	G1 TT111	G3 CFBP 6624	G1 CFBP 5771	*A. fabrum* J-07

Fig. 3. Positive and negative reactions to the 3-keto-lactose test of Bernaert and De Ley (1963) of various *A. tumefaciens* complex and *R. rhizogenes* members (A) and identification of *A. fabrum* by caffeic acid test (B). AT minimal medium (Petit et al., 1978) supplemented with caffeic acid 0.1 mg.ml⁻¹ is browning after five days of incubation except when the medium is inoculated with *A. fabrum* (here *A. fabrum* C58 or J07).

Esculin		Urease		3-ketolactose
Peptone	10 g	L-tryptophan	3 g	Lactose medium:
Esculin	1 g	Urea	20 g	Lactose 10 g
Ferric ammonium citrate	20 g	KH₂PO₄	1 g	Yeast Extract 1 g
Adjust to 1 l with H₂O		K₂HPO₄	1 g	Agar 15 g
		NaCl	5 g	Adjust volume to 1 l with H₂O
		Ethanol 95%	10 ml	
		Phenol Red	25 mg	Benedict's reagent solution A:
		Adjust to 1 l with H₂O		Sodium citrate 173 g
				Na₂CO₃ 100 g
				Adjust to 850 ml with H₂O
				Benedict's reagent solution B:
				CuSO₄ 18 g
				Adjust to 150 m with H₂O
				Add solution B to solution A

Table 3. Recipe for esculin, urease, and 3-ketolactose tests (adapted from Bernaerts and De Ley, 1963).

5. Species determination and strain typing using marker genes

5.1 Infraspecific diversity of *Agrobacterium* spp. at the genome level

Multi-locus sequence analysis (MLSA) performed with housekeeping or ecologically relevant genes *recA, mutS, gyrB, glgC, chvA, ampC*, reveals the occurrence of several strains with different alleles in most species (Table 1). As already showed with *chvA* and *recA* (Costechareyre et al., 2009, 2010), these genes allow the identification of novel isolates of *Agrobacterium* spp. at both species and strain levels. Nevertheless, even if MLSA is better than single gene analysis, we found that *recA* alone is definitively a good proxy for the identification of *Agrobacterium* species, since we have so far not found any indication that this gene has been laterally transferred between different species. Identification can be easily done by a nucleotide Blast in databases. Thus, in order to facilitate the identification of *A. tumefaciens* genomovars, we have clearly indicated the genomovar and allele variant status of sequences deposited at EMBL in the description section, as shown below.

Accession#:   FM164286
Description: Agrobacterium tumefaciens partial recA gene for recombinase A, genomovar G1, strain TT111, allele recA-G1-1

Extensive analysis of the infraspecific diversity is now conducted at the whole genome level by comparative genomics. The first analyses were done with DNA microarrays containing probes covering the whole genome of C58. This procedure permitted us to find species specific genes of *A. tumefaciens* genomovar G8 that were used to find typical phenotypic traits of this species (such as the ability to degrade the caffeic acid, Fig. 3). It was then possible to propose this genomovar as a novel species: *A. fabrum* (Lassalle et al., 2011). Comparative genomics of *Agrobacterium* spp. is now done with 22 complete genomes using the Agrobacterscope plateform at Genoscope:

https://www.genoscope.cns.fr/agc/microscope/about/collabprojects.php?P_id=51

However, while MLSA and genome sequencing are suited to type a limited number of strains for taxonomy purposes, these methods are not easily applicable to analyze large agrobacterial populations, as is required for epidemiological investigations. For this reason, we developed a PCR-RFLP approach consisting of the restriction of PCR products obtained from selected genome regions (Figure 4). The intergenic region spanning between the 16S and 23S rRNA genes (16S+ITS) was found to allow the distinction of strains within both the *R. rhizogenes* and *A. tumefaciens* complex (Ponsonnet & Nesme, 1994), with an accuracy we now know to be equivalent to those of *recA* (unpublished results).

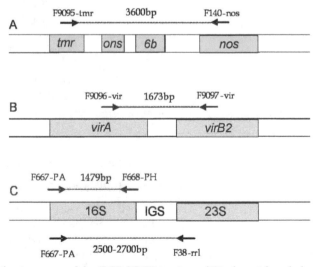

Fig. 4. Location of primers used for PCR-RLFP typing of Ti plasmid and chromosome. A, B and C indicate primer locations in the T-DNA region (regions between the *tmr* and *nos* genes), the *virA* and *virB₂* genes, and the ribosomal operon (16S and 23S rRNA genes), respectively.

5.2 Ti plasmid diversity

Although agrobacteria are primarily natural soil inhabitants and also commensal rhizospheric bacteria, they become a plant pathogen when they acquire a Ti (Tumor inducing) plasmid (size ~200 kb). Thus it becomes important to characterize agrobacterium plasmid content. Generally agrobacteria of the *A. tumefaciens* complex harbor, besides their chromosome and chromid, another large plasmid called the At plasmid (~400 kb). Large Ti and At plasmids can be visualized on an agarose gel using a method modified from Eckhardt (1978) by Wheatcroft et al. (1990), which is based on the action of gentle bacterium lysis directly in agarose gel wells in order to avoid DNA breakages. Briefly, agrobacterial strains are grown in standard medium to an optical density of 0.2 at 600 nm. After which a total of 1 ml of cell culture is centrifuged at 13,000g for 1 min at 4°C. The pellet is then washed once in 1 ml Na N-lauroyl sarcosinate 0.3% and resuspended in Ficoll buffer (10 mM Tris, pH 7.5, and 0.1 mM EDTA, pH 8.0, 20% Ficoll® 400; Sigma) and incubated on ice for 15 min. Then lysis solution consisting of 10 mM Tris, 10 mM EDTA, RNase A (0.4 mg.ml^{-1}), bromophenol red (1mg.ml^{-1}; Sigma), and lyzosyme (1mg.ml^{-1}; Sigma) is added to the cell suspension. Immediately after addition of the lysis solution to cells, 25 µl of this solution is loaded on to a 0.75% agarose gel. Migration is run for 3 h at 100V. Characteristic plasmid size of different *Agrobacterium* and *Rhizobium* strains are given figure 5. In addition, the presence of a linear chromosome (i.e. chromid) can be easily visualized in the *A. tumefaciens* complex, *A. rubi* and *A. larrymoorei* by pulse field electrophoresis (Ramirez-Bahena et al., 2011).

While pathogenic agrobacteria differ from the non-pathogenic strains by the presence of Ti or Ri plasmids, there is also a large diversity within this plasmid class that must be considered in epidemiological investigations. Briefly, the eco-pathology of crown gall and hairy root diseases is centered upon the fact that Ti and Ri plasmids both incite the diseased plant to produce a particular class of compounds called 'opines', which give the agrobacteria a unique ability to use these particular compounds for their growth. There are several kind of opines produced in tumors and consumed by agrobacteria. The most common types of Ti plasmids produce nopaline, octopine or agropine and mannopine (for a review about opines and opine concept see Dessaux et al., 1992). While genes encoding the biosynthesis and the consumption of opines are very different according to opine types, genes involved in the pathogenic process are conserved amongst Ti and Ri plasmids and can be used to define the primers to amplify Ti-plasmid specific regions. Conserved regions are the region of virulence or *vir* region involved in the processing of the DNA transferred from agrobacterium to plant, (i.e. the T-DNA) and T-DNA genes determining the tumoral morphology such as *tmr* or *tms*. Thus, the PCR-RFLP approach is also applicable to study the Ti-plasmid diversity using T-DNA and *vir* regions (Figure 4 and table 4).

There are however a diversity of Ti plasmid conserved genes that can be used to rapidly assess the opine-type of pathogenic agrobacteria. With this idea we developed a set of nopaline-type specific primers in one of the pioneering works using PCR to detect plant pathogenic bacteria (Nesme et al., 1989). Primers F14 and F44 (Table 4) allowed the amplification of a 247 bp-long fragment only with nopaline-type Ti plasmids such as pTiC58 or pTiT37, while primer set F14-F750 permits us to specifically amplify a 217 bp-long amplicon of octopine-type Ti plasmids, such as pTiB6 or pTi15955. The 'universal' primer set F14-F749 was designed based on alignments of *vir* regions of the most common Ti-plasmid opine type (Figure 6). As these primers match different genes separated by intergenic regions of different lengths, amplicon sizes can be used to determine Ti-plasmid opine types.

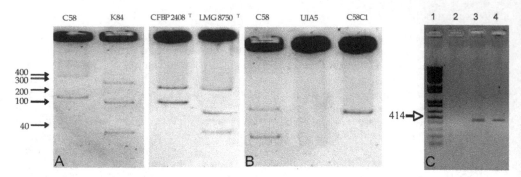

Fig. 5. Plasmids in *Agrobacterium sp.* and *Rhizobium rhizogenes*. A and B plasmid content determined by the Wheatcroft's method using gentle in well lysis of bacteria. A , *A. fabrum* C58[T], *Rhizobium rhizogenes* K84 and CFBP2408[T], *Agrobacterium vitis* LMG8750[T]. B, *A. fabrum* C58 derivatives showing At and Ti plasmids of ca. 210 and 450 kbp, respectively. Molecular weight in kb. C, PCR detection of pathogenic *Agrobacterium* in environmental DNAs with primers VCF3 and VCR3. Lane 1, marker 1kb+; 2, negative control with DNA extracted from healthy plant tissue; 3 and 4, positive detections of a *virC1-virC2* region using DNAs extracted from tumors of tomatoes infected three weeks earlier with strain C58.

Primer set				Amplified regions
PCR-RFLP typing				
F667-PA	AGAGTTTGATCCTGGCTCAG	F668-PH	AAGGAGGTGATCCAGCCGCA	*rrs* (16S rRNA)
F667-PA	AGAGTTTGATCCTGGCTCAG	F38-rrl	CCGGGTTTCCCCATTCGG	16S+ITS
F9095-tmr	CCATGTTGTTTGCTAGCCAG	F140-nos	CACCATCTCGTCCTTATTGA	T-DNA
F9096-vir	TCAAAAGGCAAGCAAGCAGATCTGG	F9097-vir	TCAGTGCCGCCACCTGCAGATTG	*virA-virB₂*
PCR detection of Ti plasmids				
F14-vir	GAACGTGTTTCAACGGTTCA	F44-vir	TGCCGCATGGCGCGTTGTAG	*virB-G* (nopaline)
F14-vir	GAACGTGTTTCAACGGTTCA	F750-vir	GTAACCTCGAAGCGTTTCAC	*virB-G* (octopine)
F14-vir	GAACGTGTTTCAACGGTTCA	F749-vir	GCTAGCTTGGAAGATCGCAC	*virB-G* (oct. + nop.)
F8521-VCF3	GGCGGGCGYGCYGAAAGRAARACYT	F8522-VCR3	CGAGATTGCGTGCTTGTAGA	*virC1-C2* (pTi & pRi)

Table 4. PCR primers used to type agrobacteria by PCR-RLFP or for the detection of Ti and Ri plasmids (adapted from Bruce et al., 1992; Ponsonnet et al., 1994 and Kawaguchi et al., 2005).

5.3 Strain and Ti plasmid diversities in crown gall outbreaks

In the case study presented below, the PCR-RFLP method was used to type agrobacteria at both chromosomal and Ti plasmid levels. Agrobacteria were isolated from tumors of hybrid poplars (*Populus alba* x *P. termuloides*) in different tree nurseries (Orléans, Loiret, France and Peyrat-le-Château, Haute-Vienne, France) supposedly affected by the same crown gall epidemic. However, results summarized in figure 7 show that the pathogenic isolates belonged to different strains (as indicated by chromosomal ribotypes) of *Rhizobium rhizogenes* (28%) or *Agrobacterium tumefaciens* (72%). For the latter taxon, strains essentially belonged to genomovar G1 (96%). Results also showed a notable diversity of Ti-plasmids that were all of the nopaline-type (data not shown), despite the fact that they were dispatched in four different PCR-RFLP patterns: pTi2516 (50%), pTiM80 (8%), pTi1903 (23%), pTi2177 (9%) and pTi292 (10%).

Rapid and Efficient Methods to Isolate, Type Strains and Determine Species of Agrobacterium spp. in Pure Culture and Complex Environments

45

```
CLUSTAL X (1.5b) multiple sequence alignment

oct   GAACGTGTTTCAACGGCTCACCTTTCAATCTAAAATC-TGAACCCTTGTTCACAGCGCTT
agr   GAACGTGTTTCAACGGCTCACCTTTCAATCTAAAATC-TGAACCCTTGTTCACAGCGCTT
nop   GAACGTGTTTCAACGGTTCACCTCTCAATCTAGGATCCTGGCCAGCCATTTGCAGCTCAA
      **************** ****** ****** ********   *** **   *      **   **** *


oct   GAGAAATTTT-CACGTGAAGGATGTAC--------------------AATCATCTCCAGCT
agr   GAGAAATTTT-CACGTGAAGGATGTAC--------------------AATCATCTCCAGCT
nop   CAGAATTTATACACGTAAAGGTTGTATTTGCTAGACTCCACTCTTTAATTTTCTCTCACT
      **** ** * ***** **** ****                     ***   ****    **


oct   AAATGGGCAGTTCGTCAGAATTGCGGCTGA-CCGCGGATGACAAAAATGCGAACCAAGTA
agr   AAATGGGCAGTTCGTCAGAATTGCGGCTGA-CCGCGGATGACGAAAATGCGAACCAAGTA
nop   ACACGGGCATTTCGGCAAGATTTCGACCAAACCGCGCACGACAGAAATGCAAACTAGATG
      * * ***** **** **  *** **  *   * ***** *  ***  ****** ***  *   *


oct   TTTCAATTTTATGACAAAAATTCTCAATCGTTGTTACAA--GTGA------AACGC----
agr   TTTCAATTTTATGACAAAAGTTCTCAATCGTTGTTACAA--GTGA------AACGC----
nop   TCTCCGTTTGATGACAAAGATTGCTGAGCATTGCTACAAACGTAATTCTACAACGCGCCA
      * **  *** ********  **    * * *** *****  ** *       *****


oct   --------TTCGA------GGTTACAGCTACTATTGATTTAGGAGATCGCCTATGGTCTC
agr   --------TTCGA------GGTTACAGCTACTATTGATTAAGGAGATCGCCTATGGTCTC
nop   TGCGGCATTTAGAAACATGGATCACAACTACTGCTGGTTAAGAAGATCGCCTATTGTCTC
      ** **     * * ***  *****   ** **  ** **  ** *********** *****


oct   GCCCCGGCGTCGTGCGTCCGCCGCGAGCCAGATCTCGCCTACTTCATAAACGTCCTCATA
agr   GCCCCGGCGTCGTGCGTCCGCCGCGAGCCAGATCTCGCCTACTTCATAAACGTCCTCATA
nop   ACCGCGCCGACGCGCATCGGCAGCGAGCCAGATTTCGCCCACCTCGTAAATGTCACCGTG
      ** **  ** ** ** **  **  *** ** *********** ***** ** **  **** ***   * *


oct   GGCACGGAATGGAATAATAACATCGATCGCCGTAGAGAGCATGTCAATCAGTGTGCGATC
agr   GGCACGGAATGGAATGATGACATCGATCGCCGTAGAGAGCATGTCAATCAGTGTGCGATC
nop   GGCACGGAAGGGTACGATGACATCAACTGCGGTTGCGAGCATGTCAATCAGGGTGCGATC
      *********  **  *   **  *****   *  **   ** **   * ***************    *******


oct   TTCCAAGCTAGC
agr   TTCCAAGCTAGC
nop   TTCCAAGCTAGC
      ************
```

Fig. 6. Aligment of *vir*G15-*vir*B11 regions amplifiable with Ti-plasmid-universal primers F14-F749 in Ti plasmids of different opine types. The first box corresponds to F14. The second box corresponds to reverse complement of F749. oct, octopine-type Ti plasmids such as pTi15955 or pTiB6. agr, agropine-mannopine-type Ti plasmid pTiBo542. nop, nopaline-type Ti plasmids such as pTiC58 or pTiT37. Amplicon lengths are 384 bp, 387 bp and 372 bp for octopine-, agropine-mannopine- and nopaline-type Ti plasmids, respectively.

Remarkably, a given Ti plasmid type could be found in different strains while a given strains could harbor different Ti plasmids, likely as the results of Ti plasmid conjugal transfers. In addition, none of the indigenous agrobacteria isolated from a non-contaminated plot harbored a Ti plasmid. Indigenous agrobacteria were found to be different of pathogenic strains isolated from tumors, except for some isolates of ribotype 2520 that harbored a pTi2516. This strongly suggests that indigenous soil agrobacteria — that were initially Ti plasmid free and consequently not pathogenic, such as 2520— were able to receive a Ti plasmid from introduced pathogenic strains, and then in turn become pathogenic. This may facilitate the establishment of Ti plasmids in indigenous soil agrobacteria and could be at the basis of perennial soil contaminations recorded for decades in several instances (Krimi et al., 2002; Costechareyre et al., 2010).

From an epidemiological point of view, the present chromosome and Ti plasmid diversities were found even in single tumors, illustrating the fact that a single isolate is not enough to characterize pathogenic populations involved in crown gall epidemics. However, by studying numerous isolates, we found that pathogenic agrobacteria were almost the same in Orléans and Peyrat-le-Château (data not shown), demonstrating that the crown gall outbreak of the latter nursery was caused by the introduction of apparently healthy (i.e. tumor free) — but in fact contaminated— plant material from Orléans. The same procedure involving both chromosome and Ti plasmid typing of numerous isolates was used to discover the source of crown gall contamination of rose bushes (Pionnat et al., 1999) and is currently used for routine epidemiological investigations.

6. Detection in complex environments

The detection of pathogenic *Agrobacterium* in complex environments without prior strain isolation can be done easily by PCR using primers designed in the conserved *vir* region (Table 4). Kawaguchi et al. (2005) also reported primers VCF3 and VCR3 amplifying DNA fragments of 414 bp in the *virC1-virC2* regions of both Ti and Ri plasmids. Ti plasmid detection can be conducted with metagenomic DNAs extracted from complex environments such as bulk soil, rhizosphere and tumors (Figure 5C). To date, in terms of accuracy and reproducibility the best results are obtained with DNA extracted with the PowerSoil® DNA isolation kit (http://www.mobio.com/soil-dna-isolation/powersoil-dna-isolation-kit.html).

In a similar manner, it is also possible to detect *Agrobacterium* and *Rhizobium* spp. in complex environments by using PCR probes designed in the 16S rDNA gene (Mougel et al., 2001). However, the 16S is not diverse enough to allow the distinction of closely related species such as *A. tumefaciens* genomovars. Instead, Nucleotide sequences of protein encoding genes have a much more resolutive power at both species and infraspecies levels. Nevertheless, the present challenge is to find gene regions well conserved within a species but with marked differences between species in order to design species specific primers. Only a small number of genes fulfill this requirement in *A. tumefaciens*, because genomovars are very closely related but also have a huge infraspecies diversity. Primary results suggest however that *recA* could be a convenient gene to find species specific primers suited for in situ detection of *Agrobacterium* species and genomovars (unpublished results).

Rapid and Efficient Methods to Isolate, Type Strains and Determine Species of Agrobacterium spp. in Pure Culture and Complex Environments

47

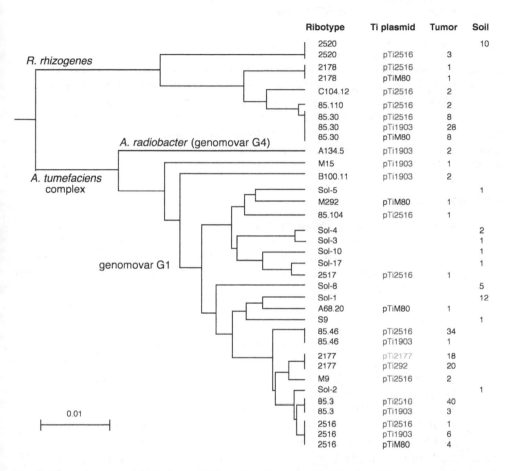

Fig. 7. Chromosome and Ti-plasmid diversity of agrobacteria isolated from poplar tumors in a single crown gall epidemic involving two distant nurseries. Ribotypes and Ti-plasmid types were determined by PCR-RFLP of the 16S+ITS region using primers F667-PA and F38 (Table 4), and digestion with *CfoI, HaeIII, NdeII* and *TaqI*; and primers F9095 and F140, and F9096 and F9097 (Table 4) and digestion with *TaqI, DdeI, MspI* and *RsaI*, respectively as described in Ponsonnet & Nesme (1994). 2520 and 2178 are ribotypes identical to those of *R. rhizogenes* strains CFBP 2520 and CFBP 2178, respectively. 2517, 2177 and 2516: ribotypes identical to those of *A. tumefaciens* genomovar-G1 strains CFBP 2517, CFBP 2177 and CFBP 2516, respectively. pTi2516, pTiM80, pTi1903, pTi2177: nopaline-type Ti-plasmids with PCR-RFLP patterns identical to those of strains M80, CFBP 1903 (i.e. C58) and CFBP 2177, respectively. pTi292: pTi2177 with the insertion element IS292 (Ponsonnet et al., 1995). Tumor indicates total numbers of isolates for each conjugated chromosome and Ti-plasmid patterns obtained in four different plots in Orléans and Peyrat-le-Château. Soil indicates agrobacteria isolated directly from a non-contaminated soil plot in the vicinity of

contaminated plots planted with poplars at Peyrat-le-Château; none of which harboring a Ti plasmid. The dendrogram was built with UPGMA using number of substitutions per nucleotidic sites calculated from RFLP patterns as genetic distance.

7. Conclusion

The genus *Agrobacterium* encompasses several species and several strains within species that are now clearly named using a novel nomenclature. A set of efficient procedures is available to study either pure culture isolates or agrobacteria directly in soil microbiomes. These methodologies are necessary to unmask the ecologies of *Agrobacterium* species that usually live sympatrically in soils. By this knowledge we expect to differentially manage agrobacterial species directly in soils to improve the crown gall as well as to stimulate their putative PGPR effects.

8. Acknowledgements

The author wishes to thank J. Puławska that kindly provided strains of novel species before the definitive publication of their results and Cyrus Mallon for reading and correcting the English. This review prepared within the framework of COST Action 873 was supported by the EcoGenome project of Agence Nationale de la Recherche (grant number ANR-08-BLAN-0090).

9. References

Beijerinck, M.W. & van Delden, A. (1902). Ueber die Assimilation des freien Stickstoffs durch Bakterien. *Zentralbl Bakteriol Parasitenk Infektionskr HygAbt*, Vol. 9, pp. 3-43

Bernaerts, M.J. & de Ley, J. (1963). A biochemical test for crown gall bacteria. *Nature*, Vol. 197, pp. 406-407, ISSN 0028-0836

Bouzar, H. & Jones, J.B. (2001). *Agrobacterium larrymoorei* sp. nov., a pathogen isolated from aerial tumours of *Ficus benjamina*. *International Journal of Systematic and Evolutionary Microbiolology*, Vol. 51, No. 3, pp. 1023–1026, ISSN 1466-5026

Bruce, K.D.; Hiorns, W.D.; Hobman, J.L.; Osborn, A.M.; Strike, P. & Ritchie, D.A. (1992). Amplification of DNA from native populations of soil bacteria by using the polymerase chain reaction. *Applied and Environmental Microbiology*, Vol. 58, No. 10, pp. 3413-3416, ISSN 0099-2240

Brisbane, P.G. & Kerr, A. (1983). Selective media for the three biovars of *Agrobacterium*. *Journal of Applied Bacteriology*, Vol. 54, No. 3, pp. 425-431, ISSN 1364-5072

Conn, H.J. (1942). Validity of the genus *Alcaligenes*. *Journal of Bacteriology*, Vol. 44, No. 3, pp. 353-360, ISSN 0021-9193

Costechareyre, D.; Bertolla, F. & Nesme, X. (2009). Homologous recombination in *Agrobacterium*: potential implications for the genomic species concept in bacteria, *Molecular Biology and Evolution*, Vol. 26, No. 1, pp. 167-176, ISSN 0737-4038

Costechareyre, D.; Rhouma, A.; Lavire, C.; Portier, P.; Chapulliot, D.; Bertolla, F.; Boubaker, A.; Dessaux, Y. & Nesme, X. (2010). Rapid and efficient identification of

Rapid and Efficient Methods to Isolate, Type Strains and Determine Species of Agrobacterium spp. in Pure Culture and Complex Environments

49

Agrobacterium species by *recA* allele analysis, *Microbial Ecology*, Vol. 60, No. 4, pp. 862-872, ISSN 0095-3628

De Ley, J. (1974). Phylogeny of prokaryotes, *Taxon*, Vol. 23, pp. 291-300, ISSN 0040-0262

De Ley, J.; Tijtgat, R.; De Smedt, J. & Michiels, M. (1973). Thermal stability of DNA: DNA Hybrids within the genus *Agrobacterium*. *Journal of General Microbiology*, Vol. 78, No. 2, pp. 241-252, ISSN 0022-1287

Dessaux, Y.; Petit, A. & Tempé, J. (1992). Opines in *Agrobacterium* biology, In : *Molecular Signals in Plant-Microbe Communications*. D. P. S. Verma, (Ed.), 109-136, CRC Press, Boca Raton

Eckhardt, T. (1978). A rapid method for the identification of plasmid desoxyribonucleic acid in bacteria. *Plasmid*, Vol. 1, No. 4, pp. 584-588, ISSN 0147-619X

Genetello, C.; Van Larebeke, N.; Holsters, M.; De Picker, A.; Van Montagu, M. & Schell, J. (1977). Ti plasmids of *Agrobacterium* as conjugative plasmids. *Nature*, Vol. 265, pp. 561-563, ISSN 0028-0836

Holmes, B. & Roberts, P. (1981). The classification, identification and nomenclature of agrobacteria. *Journal of Applied Bacteriology*, Vol. 50, pp. 443-467

Hooykaas, P.J.; Klapwijk, P.M.; Nuti, M.P.; Schilperoort, R.A. & Rörsch, A. (1977). Transfer of the *Agrobacterium tumefaciens* Ti plasmid to avirulent agrobacteria and to *Rhizobium ex planta*. *Journal of General Microbiology*, Vol. 98, No. 2, pp. 477-484, ISSN 0022-1287

Kawaguchi, A.; Inoue, K. & Nasu, H. (2005). Inhibition of crown gall formation by *Agrobacterium radiobacter* biovar 3 strains isolated from grapevine. *Journal of General Plant Pathology*, Vol. 71, No. 6, pp. 422–430, ISSN 1345-2630

Keane, P.J.; Kerr, A. & New, P.B. (1970). Crown gall of stone fruit. II. Identification and nomenclature of *Agrobacterium* isolates. *Australian Journal of Biological Sciences*, Vol. 23, No. 3, pp. 585-595, ISSN 0004-9417

Kerr, A. (1974). Soil microbiological studies on *Agrobacterium radiobacter* and biological-control of crown gall. *Soil Science*, Vol. 118, pp. 168-172, ISSN 0038-075X

Krimi, Z.; Petit, A.; Mougel, C.; Dessaux, Y. & Nesme, X. (2002). Seasonal fluctuations and long-term persistence of pathogenic populations of *Agrobacterium* spp. in soils. *Applied and Environmental Microbiology*, Vol. 68, No. 7, pp. 3358-3365, ISSN 0099-2240

Lassalle, F.; Campillo, T.; Vial, L.; Baude, J.; Costechareyre, D.; Chapulliot, D.; Shams, M.; Abrouk, D.; Lavire, C.; Oger-Desfeux, C.; Hommais, F.; Guéguen, L.; Daubin, V.; Muller, D. & Nesme, X. (2011). Genomic species are ecological species as revealed by comparative genomics in *Agrobacterium tumefaciens*. *Genome Biology and Evolution*, Vol. 3, pp. 762-781, ISSN 1759-6653

Lindström, K. &Young, J.P.W. (2011). International Committee on Systematics of Prokaryotes. Subcommittee on the taxonomy of Agrobacterium and Rhizobium: Minutes of the meeting, 7 September 2010, Geneva, Switzerland. *International Journal of Systematic and Evolutionary Microbiolology*, Vol. 61, No. 12, pp. 3089-309, ISSN: 1466-5034

Mougel, C.; Cournoyer, B. & Nesme, X. (2001). Novel tellurite-amended media and specific chromosomal and Ti plasmid probes for direct analysis of soil populations of *Agrobacterium* biovars 1 and 2. *Applied and Environmental Microbiology*, Vol. 67, No. 1, pp. 65-74, ISSN 0099-2240

Mougel, C.; Thioulouse, J.; Perrière, G. & Nesme, X. (2002). A mathematical method for determining genome divergence and species delineation using AFLP. *International Journal of Systematic and Evolutionary Microbiolology*, Vol. 52, No. 2, pp. 573-586, ISSN 1466-5026

Nesme, X.; Michel, M.-F. & Digat, B. (1987). Population heterogeneity of *Agrobacterium tumefaciens* in galls of *Populus* L. from a single nursery. *Applied and Environmental Microbiology*, Vol. 53, No. 4, pp. 655-659, ISSN 0099-2240

Nesme, X.; Leclerc, M. C. & Bardin R. (1989). PCR detection of an original endosymbiont: the Ti plasmid of *Agrobacterium tumefaciens*. In *Endocytobiology IV*, Nardon, P.; Gianinazzi-Peason, V.; Greines, A. M.; Margulis, L. & Smith, D. C. pp. 47-50, Institut National de la Recherche Agronomique, Paris

Ophel, K. & Kerr, A. (1990). *Agrobacterium vitis* sp. nov. for strains of *Agrobacterium* biovar 3 from grapevines. *International Journal of Systematic Bacteriology*, Vol. 40, No. 3, pp. 236-241, ISSN 0020-7713

Panday, D.; Schumann, P. & Das, S.K. (2011). *Rhizobium pusense* sp. nov., isolated from the rhizosphere of chickpea (*Cicer arietinum* L.). *International Journal of Systematic and Evolutionary Microbiolology*, Vol. 61, pp. 2632 -2639, ISSN 1466-5026

Panagopoulos, C.G.; Psallidas, P.G. & Alivizatos, A.S. (1978). Studies on biotype 3 of *Agrobacterium radiobacter* var. *tumefaciens*. In Station de Pathologie Végétale et Phytobactériologie, ed., Plant Pathogenic Bacteria Proc 4th Internat Conf Plant Path Bact, Angers, France, pp 221-228

Petit, A.; Tempe, J.; Kerr, A.; Holsters, M.; Van Montagu, M. & Schell, L. (1978). Substrate induction of conjugative activity of *Agrobacterium tumefaciens* Ti plasmids. *Nature*, Vol. 271, pp. 570-572, ISSN 0028-0836

Pionnat, S.; Keller, H.; Héricher, D.; Bettachini, A.; Dessaux, Y.; Nesme, X. & Poncet, C. (1999). Ti plasmids from *Agrobacterium* characterize rootstock clones that initiated a spread of crown gall disease in Mediterranean countries. *Applied Environmental Microbiology*, Vol. 65, No. 9, pp. 4197-4206, ISSN 0099-2240

Pitzscke, A. & Hirt, H. (2010). New insights into an old story: *Agrobacterium* induced tumour formation in plants by plant transformation. *EMBO Journal*, Vol. 29, No. 6, pp. 1021-1032, ISSN 0261-4189

Ponsonnet, C. & Nesme, X. (1994). Identification of *Agrobacterium* strains by PCR-RFLP analysis of pTi and chromosomal regions. *Archives of Microbiology*, Vol. 161, No. 4, pp. 300-309, ISSN 0302-8933

Ponsonnet, C.; Normand, P.; Pilate, G. & Nesme, X. (1995). IS292 : a novel insertion element from *Agrobacterium*. *Microbiology*, Vol. 141, No. 4, pp. 853-861, ISSN 1350-0872

Popoff, M.Y.; Kersters, K.; Kiredjian, M.; Miras, I. & Coynault. C. (1984). Position systématique de souches de *Agrobacterium* d'origine hospitalière. *Annals of Microbiology*, Vol. 135, No. 3, pp. 427-442, ISSN 1590-4261

Portier, P.; Fisher-Le Saux, M.; Mougel, C.; Lerondelle, C.; Chapulliot, D.; Thioulouse, J. & Nesme, X. (2006). Identification of genomic species in *Agrobacterium* biovar 1 by AFLP genomic markers. *Applied and Environmental Microbiology*, Vol. 72, No. 11, pp. 7123-7131, ISSN 0099-2240

Puławska, J.; Willems, A. & Sobiczewski, P. (2012) *Rhizobium skierniewicense* sp. nov. isolated from tumors on chrysanthemum and *Prunus* in Poland, *International Journal of Systematic and Evolutionary Microbiology* doi: 10.1099/ijs.0.032532-0 (in press), ISSN 1466-5026

Puławska, J.; Willems, A.; De Meyer, S. & Süle, S. *Rhizobium nepotum* sp nov. isolated from tumors on different plant species, unpublished paper

Ramirez-Bahena, M.; Nesme, X. & Muller, D. (2012). Rapid and simultaneous detection of linear chromosome and large plasmids in Proteobacteria. *Journal of basic microbiology*, Vol. 52, No. 1, pp. 1-4 ISSN 1521-4028

Riker, A.J.; Banfield, W.M.; Wright, W.H.; Keitt, G.W. & Sagen, H.E. (1930). Studies on infectious hairy root of nursery trees of apples. *Journal of Agricultural Research*, Vol. 41, pp. 507-540

Roy, M.A. & Sasser, M. (1983). A medium selective for *Agrobacterium tumefaciens* biotype 3. *Phytopathology*, Vol. 73, No. 1, pp. 810, ISSN 0031-949X

Smith, E.F. & Townsend, C.O. (1907). A plant-tumor of bacterial origin. *Science*, Vol. 25, pp. 671-673, ISSN 0036-8075

Stackebrandt, E.; Frederiksen, W.; Garrity, G.M., Grimont, P.A.; Kampfer, P.; Maiden, M.C.; Nesme, X.; Rossello-Mora, R.; Swings, J.; Truper, H.G.; Vauterin, L.; Ward, AC. & Whitman, W.B. (2002). Report of the ad hoc committee for the re-evaluation of the species definition in bacteriology. *International Journal of Systematic and Evolutionary Microbiology*, Vol. 52, No. 3, pp. 1043-1047, ISSN 1466-5026

Van Larebeke, N.; Engler, G.; Holsters, M.; Van den Elsacker, S.; Zaenen, I.; Schilperoort, R.A. & Schell, J. (1974). Large plasmid in *Agrobacterium tumefaciens* essential for Crown Gall-inducing ability. *Nature*, Vol. 252, pp. 169-170, ISSN 0028-0836

Vogel, J.; Normand, P.; Thioulouse, J.; Nesme, X. & Grundmann, G. (2003). Relationship between spatial and genetic distance in *Agrobacterium* spp. in 1 cubic centimeter of soil. *Applied and Environmental Microbiology*, Vol. 69, No. 3, pp. 1482-1487, ISSN 0099-2240

Waine, L.G.; Brenner, D.J.; Colwell, R.R.; Grimont, P.A.D.; Kandler, O.; Krichevsky, M.I.; Moore, L.H.; Moore, W.E.C.; Murray, R.G.E.; Stackebrandt, E.; Starr, M.P. & Trüper, H.G. (1987). Report of the ad hoc committee on the reconciliation of approaches to bacterial systematics. *International Journal of Systematic and Evolutionary Microbiology*, Vol. 37, No. 4, pp. 463-464, ISSN 1466-5026

Wheatcroft, R.; McRae, D.G. & Miller, R. W. (1990). Changes in the *Rhizobium meliloti* genome and the ability to detect supercoiled plasmids during bacteroid development. *Molecular Plant-Microbe Interactions*, Vol. 3, No. 1, pp. 9-17, ISSN 0894-0282

Young, J.M.; Kuykendall, L.D.; Martinez-Romero, E.; Kerr, A. & Sawada, H. (2001). A revision of *Rhizobium* Franck 1889, with an emended description of the genus, and the inclusion of all species of *Agrobacterium* Conn 1942 and *Allorhizobium undicola* de Lajudie et al. 1998 as new combinations: *Rhizobium radiobacter, R. rhizogenes, R. rubi, R. undicola* and *R. vitis. International Journal of Systematic and Evolutionary Microbiology*, Vol. 51, No. 1, pp. 89-103, ISSN 1466-5026

Young, J.M.; Pennycook, S.R. & Watson, D.R.W. (2006). Proposal that *Agrobacterium radiobacter* has priority over *Agrobacterium tumefaciens*. Request for an Opinion. *International Journal of Systematic and Evolutionary Microbiology*, Vol. 56, No. 2, pp. 491-493, ISSN 1466-5026

Part 2

Genetic Modified Organisms in Today's Agriculture

3

Testing Methods for Agriculture and Food Safety

Jose C. Jimenez-Lopez[1*] and María C. Hernandez-Soriano[2]
[1]Purdue University, Department of Biological Sciences,
College of Science, West Lafayette, IN
[2]North Carolina State University, Department of Soil Science,
College of Agriculture and Life Sciences, Raleigh, NC
USA

1. Introduction

Sustainable agricultural production is a critical concern in response to global climate change and population increase (Brown and Funk, 2008; Turner et al., 2009). In addition, recent increased demand for Biofuel crops has created a new market for agricultural products. One potential solution is to increase plant yield by designing plants based on a molecular understanding of gene function and on the regulatory networks involved in stress tolerance, development and growth (Takeda & Matsuoka, 2008). Recent progress in plant genomics, as well as sequencing of the whole genome of new plant species have allowed to isolate important genes, discover new traits and to analyze functions that regulate yields and tolerance to environmental stress.

Recent remarkable innovations in platforms for omics-based research (genomic, proteomic, transcriptomic, metabolomic) and application development provide crucial resources to promote research in model and applied plant species. A combinatorial approach using multiple omics platforms and integration of their outcomes is now an effective strategy for clarifying molecular systems integral to improving plant productivity. Furthermore, support of comparative genomics and proteomis among model and applied plants allow to reveal the biological properties of each species and to accelerate gene discovery and functional analyses of genes. Bioinformatics platforms and their associated databases are also essential for the effective design of approaches making the best use of genomic resources, including resource integration.

Currently, many crop species can be considered important on a global scale for food security in which, have been developed large-scale genomic and genetic resources, i.e. array technology markers, expressed sequence tags or transcript reads, bacterial artificial chromosome libraries, genetic and physical maps, and germplasm stocks with rich genetic diversity, among others. These resources have the potential to accelerate gene discovery and initiate molecular breeding in these crops, thereby enhancing crop productivity to ensure food security in developing countries. In this line, and with all above molecular genetics

---

* Corresponding Author

tools development, it has been possible that more and more plant crop species has been modified genetically obtaining desirable traits compared with wild plants.

Classical plant breeding uses deliberate interbreeding (crossing) of closely or distantly related individuals to produce new crop varieties or lines with desirable properties. Plants are crossbred to introduce traits/genes from one variety or line into a new genetic background. Progeny from the cross would then be crossed with the high-yielding parent to ensure that the progeny were most like the high-yielding parent, (backcrossing). The progeny from that cross would then be tested for desirable traits, i.e. resistance and high-yielding resistant plants to be further developed. Plants may also be crossed with themselves to produce inbred varieties for breeding.

Classical breeding relies largely on homologous recombination between chromosomes to generate genetic diversity (Zhu et al., 2009). The classical plant breeder may also makes use of a number of in vitro techniques such as protoplast fusion, embryo rescue or mutagenesis to generate diversity and produce hybrid plants that would not exist in nature.

Traits that breeders have tried to incorporate into crop plants in the last 100 years include:

1) Increased quality and yield of the crop, 2) Increased tolerance of environmental pressures (salinity, extreme temperature, drought), 3) Resistance to viruses, fungi and bacteria, 4) Increased tolerance to insect pests, and 5) Increased tolerance of herbicides.

Modern plant breeding uses techniques of molecular biology to select, or in the case of genetic modification, to insert, desirable traits into plants (Bhatnagar-Mathur et al., 2008). Genetic modification of plants is achieved by adding a specific gene or genes to a plant, or by knocking out a gene with RNAi, to produce a desirable phenotype. Genetic modification can produce a plant with the desired trait or traits faster than classical breeding because the majority of the plant's genome is not altered. The use of tools such as molecular markers or DNA fingerprinting can map thousands of genes, allowing plant breeders to screen large populations of plants for those that possess the trait of interest. The screening is based on the presence or absence of a certain gene as determined by laboratory procedures, rather than on the visual identification of the expressed trait in the plant. The majority of commercially released transgenic plants are currently limited to plants that have introduced resistance to insect pests and herbicides. Insect resistance is achieved through incorporation of a gene from Bacillus thuringiensis (Bt) that encodes a protein that is toxic to some insects. For example, the cotton bollworm, a common cotton pest, feeds on Bt cotton it will ingest the toxin and die. Herbicides usually work by binding to certain plant enzymes and inhibiting their action. The enzymes that the herbicide inhibits are known as the herbicides target site. Herbicide resistance can be engineered into crops by expressing a version of target site protein that is not inhibited by the herbicide. This is the method used to produce glyphosate resistant crop plants (Pollegioni et al., 2011).

Testing for the presence of agricultural biotechnology products is being performed on many grain and food products. Currently, there is an absence of standardized tests to detect genetic modified (GM) crops, which can result in inaccurate claims and enforcement actions being taken without a means to challenge the results. Development of reliable, validated methods is necessary to avoid negative economic impacts due in invalid test results, as well as to ensure the safety of the consumer under arising of GM product in the markets. We also emphasize the need for such global compatibility of test results in order to facilitate international trade. A quality control of GM-crops product should be implemented through self, supervisory, and peer review systems, which have to be organizationally independent of the testing staff or organization, performing internal test in accordance with the government requirements for.

Identification of GM product in the markets is another pending task, due that government from different countries are not in agreement about information that labelling product should contain. Authoritative, independent and public acceptable of green (eco-)/GM label scheme that identify products no-GM, in many case more environmentally desirable than other similar GM products with the same function in the market, is urged to be implemented with a general acceptance. Establishing compliance with GM food labeling laws is dependent on the availability of test methods capable of determining the presence and or concentration of GM ingredients in food or bulk consignments of agricultural commodities such as seed and grain.

Thus, current global regulatory requirements for labeling of products derived from plant biotechnology means that test methods, i.e. for the introduced trait(s) have to be developed and validated. Such methods require appropriate reference materials, controls and protocols in order to give accurate and precise identification and quantification of GM products.

2. Genetic modified crops

The term genetically-modified (GM) foods u organisms (GMOs) is most commonly used to refer to crop plants created for human or animal consumption, using molecular biology techniques. These plants have been modified in the laboratory to enhance desired traits such as increased resistance to herbicides or improved nutritional content. The enhancement of desired traits has traditionally been undertaken through breeding, but conventional plant breeding methods can be very time consuming and are often not very accurate.

During the past decade, a large number of genetically modified (GM) crops have been developed using methods of modern biotechnology. These GM or "biotech" crops exhibit unique agronomic traits such as herbicide tolerance or insect resistance, which offer significant benefits to farmers. The development of GM crops is accomplished by using molecular biology methods, essentially by the integration of novel DNA sequences into the plant genome. The new DNA encodes for the expression of the novel protein in the targeted tissue, resulting in the unique agronomic trait. The novel protein and DNA are present in many parts of the plant, in harvested grain, and often in the food fractions prepared from grain.

The production and global trade of genetically modified (GM) grain is increasing. At the same time, companies are required to provide validated diagnostic methods proving the inclusion of GM material, as a condition of the regulatory approval process in some jurisdictions. Numerous governmental agencies and industry organizations are attempting to develop standardization guidelines independently. Global harmonization of these efforts is necessary to ensure a consistent standard. An international coordination of detection methods for plant biotechnology products and the proper development of guidelines for their use are necessary and an awaiting mission.

2.1 Detection methods for GM-Crops

The detection of genetically modified organisms in food or feed is possible by using well-developed genetic and biochemical tools. It can either be qualitative, showing which genetically modified organism (GMO) is present, or quantitative, measuring in which amount a certain GMO is present. Being able to detect a GMO is an important part of food safety, as without detection methods the traceability of GMOs would rely solely on documentation.

2.1.1 Methods based on DNA: polymerase chain reaction

Methods for GMO should contain three analytical components of tests for detecting the presence of transgenic plant products: (1) detection, to screen for the presence of GM events in food and agricultural products; (2) identification, to reveal how many GM events are present and determine their molecular registers; and (3) quantification, to determine the amount of authorized GM product and compliance with threshold regulation. Analytical procedures for GM plants are directed to detect either the novel gene product or the gene construct itself.

Testing on GMOs in food and feed is routinely done by molecular techniques like DNA microarrays or qPCR, and the test can be based on screening elements (like p35S or terminator Nos) or specific markers for the official GMOs (like Bt11 or GT73).

Technologies involving amplification of gene fragments by the polymerase chain reaction (PCR) have the greatest potential for detecting transgenic plants and foodstuffs derived from them (Ahmed, 2002; Schmidt et al., 2008). PCR is a biochemistry and molecular biology technique for isolating and exponentially amplifying a fragment of DNA, via enzymatic replication, without using a living organism. It enables the detection of specific strands of DNA by making millions of copies of a target genetic sequence. Target sequences may be those of the marker gene, the promoter, the terminator, or the transgenes themselves. Confirmatory tests are essential to ensure authenticity of PCR product. The use of genomic fragments that include the border sequence at the insertion site and the inserted genes (edge or junction fragments) may be a better target for unequivocal identification of GM plant sources (Windels et al., 2001), particularly when detection is based on regulatory sequences in promoters and terminators that could occur in microbial contaminants.

Improving PCR based detection of GMOs is a further goal of different governmental research programme. Research is now underway to develop multiplex PCR methods that can simultaneously detect many different transgenic lines. Another major challenge is the increasing prevalence of transgenic crops with stacked traits. This refers to transgenic cultivars derived from crosses between transgenic parent lines, combining the transgenic traits of both parents.

Whether or not a GMO is present in a sample can be tested by qPCR, but also by multiplex PCR. Multiplex PCR uses multiple, unique primer sets within a single PCR reaction to produce amplicons of varying sizes specific to different DNA sequences. By targeting multiple genes at once, additional information may be gained from a single test run that otherwise would require several times the reagents and more time to perform. Furthermore, DNA array technology using chip platforms may also be a useful tool for determining the presence of insertional sequences. The possibility of testing against a large number of oligonucleotides representing various gene sequences will be particularly useful when the specific construct is unknown. To avoid any kind of false positive or false negative testing outcome, comprehensive controls for every step of the process is mandatory. A CaMV check is important to avoid false positive outcomes based on virus contamination of the sample.

Sometime it is required to quantify GM product that contain food or directly in plants. For that purpose it is frequently used quantitative PCR (qPCR), to measure amounts of transgene DNA or PCR product in a food or feed sample, preferably real-time qRT-PCR (Ref.), with currently the highest level of accuracy. If the targeted genetic sequence is unique to a certain GMO, a positive PCR test proves that the GMO is present in the sample.

In addition, the array-based methods combine multiplex PCR and array technology to screen samples for different potential GMOs (Querci et al., 2009; Dorries et al., 2010) combining different approaches (screening elements, plant-specific markers, and event-specific markers). Advanced PCR technologies, including competitive multiplex PCR and real-time PCR, are useful for quantifying the level of GM plant material in foodstuffs (Matsuoka et al., 2000).

Alternatively, there are methods to detect specific genetic construct, instead the specific genetic product. Since different GMOs may produce the same protein, construct-specific detection can test a sample for several GMOs in one step, but is unable to tell precisely, which of the similar GMOs are present.

Almost all transgenic plants contain a few common building blocks that make unknown GMOs easier to find. Even though detecting a novel gene in a GMO can be like finding a needle in a haystack, the fact that the needles are usually similar makes it much easier. Researchers now compile a set of genetic sequences characteristic of GMOs. After genetic elements characteristic of GMOs are selected, methods and tools are developed for detecting them in test samples. Approaches being considered include microarrays and anchor PCR profiling.

2.1.2 Methods based on protein: immunoassay technology

GM content can be determined by methods that detect either the novel protein or the inserted DNA. Detection of the novel proteins produced by GM crops relies almost exclusively on the application of immunoassay technology (Fantozzi et al., 2007). Commercial immunoassays are available for most of the GM crops on the market today and have been used in a variety of large-scale applications, determining GM content (%GM) ensuring compliance with non-GM labeling requirements, and confirming the presence of high-value commodities.

Immunoassays are based on the reaction of an antigen (Ag), e.g., transgenic protein, with a specific antibody (Ab) to give a product (Ag-Ab complex) that can be measured. There are many different immunoassay formats, and the choice of format is dependent on the target molecule and application. For macromolecules, the most commonly used test formats are enzyme-linked immunosorbent assay (ELISA) that can be used as either a qualitative or a quantitative assay, and lateral flow device (LFD) designed for qualitative yes/no testing.

ELISA based commercial kits are available for serological detection of selected GM gene products. ELISA is a comparatively easy and cost-effective procedure to apply to large numbers of samples, but specificity of antibodies is critical for an accurate test.

Two other test formats used for seed quality testing are Western blot and immunohistochemical staining.

LFDs are used for qualitative or semiquantitative detection of antigens. LFDs for the detection of GM proteins use antibodies in the same sandwich immunoassay format used in ELISA, except that the secondary antibody is labeled with a colored particle such as colloidal gold rather than an enzyme as a means of generating a visible signal.

Furthermore, the Western blot is primarily a qualitative analytical method and is particularly useful in protein characterization because it provides additional information regarding molecular weight. Immunohistochemical staining is used to determine the location of the expressed proteins in the plant.

The key component of immunoassay, antibody, have the atribute that makes it useful as a reagent in a diagnostic kit, being its capacity to bind specifically and with high affinity to the antigen that elicited its production. Polyclonal antibodies are relatively easy and inexpensive to prepare in a relatively short time frame (e.g., 3–4 months); however, the quality of the antibody reagent varies from animal to animal, and it is necessary to prepare large pools of qualified reagent to support long-term commercial production of uniform product. Monoclonal antibodies require greater time (e.g., 6 months) and skill to produce and are more expensive to develop than polyclonal antibodies. In applications where discrimination between very closely related molecules is required, it may be more advantageous to use a highly specific monoclonal antibody reagent. Conversely, in an application designed to detect all the members of a family of closely related molecules it may be more advantageous to use a polyclonal antibody reagent. The selection of one reagent type over another is dependent on the desired performance characteristics of the test method.

Another key component of an immunoassay is the antigen that can be defined as substances that induce a specific immune response resulting in production of antibodies. The interaction between antibody and antigen involves binding of the antigenic epitope to the complementarily determining region (CDR) of the antibody. The strength of binding between the 2 is referred to as the affinity of the bond. In general, the greater the affinity of the bond, the greater the sensitivity (lower limit of detection; LOD) of the test method. Sensitivity of a test method is determined not only by the affinity of the antibody for the antigen, but by factors such as protein expression level, extraction efficiency, and the size of the sample taken for analysis. In addition, an antibody binds only to the antigenic determinant that elicited its production. This specificity enables the development of test methods that require minimal sample preparation. Cross-reactivity can result in false-positive responses over-estimation of antigen concentrations. Cross-reactivity of an antibody to a component of the sample or other GM crop is highly unlikely and almost never a significant issue.

The ideal antigen for immunization would be the actual GM protein as it is expressed in the plant. However, purification of the novel protein from plant tissue can be difficult and may result in undesirable modifications to the target protein. In addition, purification rarely results in 100% pure protein and immunization of animals with such preparations results not only in the production of antibodies to the target protein but to the contaminants as well. Polyclonal antibodies made from these preparations typically exhibit high background and poor sensitivity. A more common approach to making antibodies to GM proteins is to express and purify the protein of interest from an alternate host such as *E. coli* using genetic engineering techniques. Although the amino acid sequence of these recombinant proteins may be the same as the plant-produced protein, post-translational modification may be subtly different, and purification may result in modifications to the secondary and tertiary structure (e.g., denaturation). As long as antibodies that bind to the plant-produced protein with sufficient sensitivity and specificity can be isolated, then differences in structure between plant-produced and microbial-derived proteins are not an issue.

In certain instances where purified or recombinant antigens are not available or are exceedingly difficult to obtain, or where antibodies to very specific amino acids are desired, short peptides conjugated to carrier proteins may be used to develop antibodies. However, peptide antibodies may be more reactive to denatured forms of the protein and therefore often find better utility in Western blot (De Boer, 2003; Grothaus et al., 2006).

2.1.3 Others methods

To date there is no molecular approach to testing that can distinguish between the presence of a low (or high) percentage of GM events in a bulk sample, and the presence of a mixture of the two or more individual events that comprise the stack. In addition, there are added problems because gene products can become degraded during preparation and cooking. Thus, serology may not be suitable for their detection in some processed food products.

These reasons have lead to use alternative methods, por example to monitor changes in the chemical profile of oils derived from GM Plants by using chromatophaphic methods, i.e. HPLC (Lopez et al., 2009), or near infrared spectroscopy (NIR) for detection of changes in fibre structure (Michelini et al., 2008).

NIR detection is a method that can reveal what kinds of chemicals are present in a sample based on their physical properties. It is not yet known if the differences between GMOs and conventional plants are large enough to detect with NIR imaging. Although the technique would require advanced machinery and data processing tools, a non-chemical approach could have some advantages such as lower costs and enhanced speed and mobility.

Another alternative method to differentiate and quantify GM from non GM seed contained in a seed lot is a statistical approach. The approach is a pooled testing approach and involves the examination of as many as 10-20 pools. However, if the percentage of positive seeds in the sample is higher than a few percent of the seeds, the model may not give clear results.

2.1.4 Reference materials, standards and control for validation and standardization of detection and analysis methods

The request for powerful analytical methods for routine detection of GMOs by accredited laboratories has called attention to international validation and preparation of official and non-commercial guidelines. Among these guidelines are preparations of certified reference material (CRM), sampling, treatment of samples, production of stringent analytical protocols, and extensive ring-trials for determination of the efficacy of selected GMO detection procedures. Any detection method to be implemented in the identification of GM crop plants and its derived products used in food needs materials to be used for calibration and validation of such detection methods as well as proficiency testing of laboratories. The reference materials should be controlled and regulated by government agencies as general use, in order to ensure a globally harmonized approach and provide them under principles for transfer in order to control the distribution and use of intellectual property (Trapmann et al., 2010).

Reference material is material with sufficiently stable and homogeneous properties and well established to be used for calibration, the assessment of a measurement method or for assigning values to materials. Certified reference material (CRM) is reference material accompanied by a certificate issued by a recognised body indicating the value of one or more properties and their uncertainty. The certified values of these materials have been established during the course of a certification campaign including inter-laboratory studies (which should be available upon request). In the absence of CRM, standards validated by a laboratory can be used.

Reference Materials are required as reference standards in method calibration and must be produced according to international standards and guidelines and may be certified. Reference materials will be made available for all products which are commercially available. These reference materials will be made available globally and on a single GM event detection, and designated by a third-party source. This source will be selected by each company based upon factors such as global presence, operational independence or experience in working with such materials under ISO standards. For compliance with the 1% threshold level of cross-contamination of unmodified foods with GM food products, certified reference materials for precise quantification and method validation are needed. According to commonly accepted rules, the production of reference materials should preferably follow metrological principles and should be traceable to the SI system. Arbitrary definition of measurement units could lead, as a consequence, to difficulties with non-consistent standards and a lack of long-term reproducibility. In the future, efforts should be concentrated on establishing reliable quantification methods accompanied by the production of reference materials with high DNA quality and DNA degraded under controlled conditions (simulating real samples in food production) using very well characterized base materials.

European Union's Joint Research Centre, Institute for Reference Materials and Measurements, Belgium, is currently developing a system for distribution of GMO reference material (http://www.irmm.jrc.be/).

Identification and quantification of gene products in a GM plant must be done with standards that correlate to known concentrations of the antigen (protein) that it is used to produce a dose-response curve. The standard curve and the assay response from the samples are used to determine the antigen concentration. The material used to make the standards should yield a response that correlates to the actual concentration of antigen in the sample type and assay conditions specified by the test procedure. Recombinant proteins, which contain a similar or identical amino acid sequence and immunoreactivity as the GM plant-expressed protein, are often used as ELISA standards. Uniform preparations of actual samples (such as ground corn) having known concentrations of GM proteins may also be used as standards (Trapmann et al., 2002). Protein reference materials are critical for the validation of externally operated immunochemistry processes. Reference materials can be derived from a number of production sources, and can take on a variety of final forms (stabilized plant extracts to highly pure protein). Three types of certified GMO reference samples for GMO testing are especially needed: 1) DNA-CRM, 2) matrix-CRM for events of major importance, and 3) protein-CRM. An important issue to consider is that the CRMs are stable and non-degraded. Often problems with degradation of CRMs are encountered.

The European Network of GMO Laboratories has prepared a list of wishes concerning CRMs for GMO inspection as follows:

1. For production of GMO-CRMs one variety per transformation event common in USA and EU should be used.
2. GMO and non GMO should be corresponding near isogenic lines.
3. For each EU-approved GMO varieties CRMs are needed (T25, Bt176, Mon810, Bt11, Mon809, Ms1/Rf1, Ms3/Rs8, RR soy, topas 19/2).
4. CRMs for some special cases of US-approved lines like CBH351 CRMs should be available.
5. Powdery reference materials from certified commercial seeds for relative quantification of GMO should be available with a GMO content of 100, 5, 2, 1, 0.1, and 0 %.
6. Plasmids would be helpful for absolute quantification (native and competitors sequence, transgenic GMO, and housekeeping sequences).

Controls are reagents and specifications that validate each method run. Reagent controls may be different from standards. Per example, every ELISA test, qualitative or quantitative, should include known positive and negative controls to ensure assay validity. Typical controls specify limits for background, assay response to a known concentration, quantitative range, and variability between replicates.

Validation of methods is the process of showing that the combined procedures of sample extraction, preparation, and analysis will yield acceptably accurate and reproducible results for a given analysis in a specified matrix. For validation of an analytical method, the testing objective must be defined and performance characteristics must be demonstrated. Performance characteristics include accuracy, extraction efficiency, precision, reproducibility, sensitivity, specificity, and robustness. The use of validated methods is important to assure acceptance of results produced by analytical laboratories.

Each new method should be tested in trials using numerous laboratories in order to demonstrate reproducible, sensitive and specific results. In these trials the same measurements should be assessed on identical materials. The experimental designs of each trial are crucial and several questions should be considered when planning such experiments. Examples of important issues to consider include availability of satisfactory standards, number of laboratories and how they should be recruited. It is also necessary to specify the manner of calculating and expressing test result.

Unfortunally, at this moment, no single validated method has yet been developed which is capable of accurately determining all GM products in a timely and cost effective manner. Testing programs will need to incorporate the best qualities of each technology in developing testing programs. The collaborative efforts of many organizations will be required to facilitate the development of reliable, validated diagnostic tests with broad global acceptance among users and regulators.

According to European Union legislation state laboratories participating in inspection should, whenever possible, use validated analytical methods. This is also the case for all laboratories aiming at accreditation. There are some examples of methods that have been validated or accredited recently are given below:

1. Bt176, Bt11, T25 and MON810 maize using real time quantitative PCR have been accomplished by the BgVV, Federal Institute for Health protection of Consumers and Veterinary Medicine in Germany).
2. A PCR and an ELISA method for Roundup Ready™ soybean and a PCR for Maximizer maize (Bt176) have been validated for commercial testing of grain by the European Union's Joint Research Centre, JRC.
3. An ELISA for MON810 maize has also been validated by AACC (American Association of Cereal Chemists).
4. The Varietal ID PCR methods (based on primers that span unique sequence junctions) have been accredited through the United Kingdom Accreditation System (UKAS).

However, these methods based on a relatively expensive instrumentation, requiring substantial efforts in training and available only to a limited number of participants as e.g. Real-time PCR or Microarrays for validation studies may not be useful at the moment, as methods to be implemented in routine laboratories on European scale. Furthermore, GMO testing laboratories should participate in an internationally recognised external quality control assessment and accreditation scheme. In accordance with this, authorised laboratories (approved for official inspection purposes) must participate regularly in appropriate proficiency testing schemes.

2.2 Uses and concerns about genetically modified organisms

GMOs are used in biological and medical research, production of pharmaceutical drugs, experimental medicine (e.g. gene therapy), and agriculture (e.g. golden rice). The term "genetically modified organism" does not always imply, but can include, targeted insertions of genes from one species into another.

To date the most controversial but also the most widely adopted application of GMO technology is patent-protected food crops which are resistant to commercial herbicides or are able to produce pesticidal proteins from within the plant, or stacked trait seeds, which do both. Transgenic animals are also becoming useful commercially. On February 6, 2009 the U.S. Food and Drug Administration approved the first human biological drug produced from such an animal, a goat. The drug, ATryn, is an anticoagulant which reduces the probability of blood clots during surgery or childbirth. It is extracted from the goat's milk (Niemann & Kues, 2007). Furthermore, transgenic plants have been engineered to possess several desirable traits, such as resistance to pests, herbicides, or harsh environmental conditions, improved product shelf life, and increased nutritional value. Since the first commercial cultivation of genetically modified plants in 1996, they have been modified to be tolerant to the herbicides glufosinate and glyphosate, to be resistant to virus damage as in Ring-spot virus-resistant GM papaya, grown in Hawaii, and to produce the Bt toxin, an insecticide that is non-toxic to mammals (Nasiruddin & Nasim, 2007).

Most GM crops grown today have been modified with "input traits", which provide benefits mainly to farmers. The GM oilseed crops on the market today offer improved oil profiles for processing or healthier edible oils (Sayanova & Napier, 2011). The GM crops in development offer a wider array of environmental and consumer benefits such as nutritional enhancement, drought and stress tolerance. Other examples include a genetically modified sweet potato, enhanced with protein and other nutrients, while golden rice, developed by the International Rice Research Institute (IRRI), has been discussed as a possible cure for Vitamin A deficiency.

The most common genetically engineered (GE) crops now being grown are transgenic varieties of soybean, canola, cotton, and corn. Varieties of each of these crops have been engineered to have either herbicide tolerance or insect resistance (or in a few cases, both). All of the genetically engineered insect-resistant crop varieties produced so far use specific genes taken from *Bacillus thuringiensis*, a common soil bacterium, to produce proteins that are toxic to certain groups of insects that feed on them. Currently, only Bt corn and Bt cotton varieties are being grown in the U.S., but Bt potatoes were on the market for several years until being discontinued in 2001. In addition, several different genetic modifications have been used to engineer tolerance to herbicides, the most widely adopted GE trait overall. Genetically engineered herbicide tolerant varieties of each of the four major crops listed above have been developed for use with glyphosate or glufosinate herbicides, and some cotton varieties grown in the U.S. have genetically engineered tolerance to bromoxynil or sulfonylurea herbicides. About half of the papaya crop produced in Hawaii is now from genetically engineered virus-resistant varieties, but most of the world-wide papaya crop is not genetically engineered. There is currently some limited production of squash genetically engineered for virus resistance in the U.S.

All together, about 50 different kinds of genetically engineered plants (each developed from a unique "transformation event") have been approved for commercial production in the U.S. These include 12 different crops modified to have six general kinds of traits:

Transgenic trait	Crops
Insect resistance	Corn, Cotton, Potato, Tomato
Herbicide tolerance	Corn, Soybean, Cotton, Canola, Sugarbeet, Rice, Flax
Virus resistance	Papaya, Squash, Potato
Altered oil composition	Canola, Soybean
Delayed fruit ripening	Tomato
Male sterility and restorer system (used to facilitate plant breeding)	Chicory, Corn

Table 1. Genetically modified crops.

Not all of the genetically engineered varieties that have received regulatory approval are currently being grown. Some have not yet been marketed (herbicide tolerant sugarbeets and most kinds of GE tomatoes, for example), and some have been commercially grown but were later withdrawn from the market. More details on the transgenic crops listed in the table above and short descriptions of how each of the transgenic traits works are available at http://www.comm.cornell.edu/gmo/traits/traits.html.

There are many perceived risks and benefits associated with the use of transgenic crop plants for agricultural food production (Wolfenbarger & Phifer, 2000). Some of the risks relate to the use of specific transgenes while others emphasize a broader concern that addresses the entire approach of engineering heterologous genes into plants.

Most concerns about GM foods fall into three categories: 1) *environmental hazards*, including an unintended harm to other organisms, i.e. B.t. corn caused high mortality rates in monarch butterfly caterpillars; reduction of the effectiveness of pesticides; and gene transfer to non-target species, i.e. transfer of the herbicide resistance genes from the crops into the weeds.

2) *Human health risks*, including Allergenicity and toxic effects and 3) *Economic concerns*, since GM food production is a lengthy and costly process, being many new plants GM technologies patented raising the price of seeds.

One of the primary concerns about genetically engineered crop plants is that they will hybridize with wild relatives, permitting the transgene to escape and spread into the environment, which depends on its potential fitness impact. Depending on the nature of the plant and its propensity to cross-pollinate, the flow of transgenes in the field may be an important consideration. Perhaps in some instances, transgenic plants should not be grown in geographic areas where close relatives may be pollinated with transgene containing pollen. It is feared that gene flow to non target plant populations may diminish diversity within plant species. Whether or not genetic diversity is threatened by gene flow when transgenic plants are grown in the vicinity of a native gene pool, the susceptibility of native flora to contamination by transgenes ought to be taken into account.

There are several possible solutions to the problems mentioned above. Genes are exchanged between plants via pollen. Two ways to ensure that non target species will not receive introduced genes from GM plants are to create GM plants that are male sterile (do not produce pollen) or to modify the GM plant so that the pollen does not contain the introduced gene (Warwick et al., 2009). Cross-pollination would not occur, and if harmless insects such as Monarch caterpillars were to eat pollen from GM plants, the caterpillars would survive.

Another possible solution is to create buffer zones around fields of GM crops. Beneficial or harmless insects would have a refuge in the non GM corn, and insect pests could be allowed to destroy the non GM corn and would not develop resistance to Bt pesticides. Gene transfer to weeds and other crops would not occur because the wind blown pollen would not travel beyond the buffer zone (Hüsken & Dietz-Pfeilstetter, 2007).

Potential unanticipated events relating to the safety and acceptability of transgenic plants include the transfer of antibiotic-resistant genes, up regulation of non-target genes by foreign promoter sequences, production of allergenic compounds and proteins, including cross-reactivity between plant-derived food and/or pollen (Jimenez-Lopez et al. 2011) , or gene products with mammalian toxicity. Different strategies have been developed for reducing the probability and impact of gene flow, including physical separation from wild relatives and genetic engineering. Mathematical models and empirical experimental evidence suggest that genetic approaches have the potential to effectively prevent transgenes from incorporating into wild relatives and becoming established in wild populations that are not reproductively isolated from genetically engineered crops. In addition, transgene strategies for controlling plant disease do not raise some of the same concerns that relate to the release of herbicide-tolerant cultivars or insect-protected varieties. The environmental and food safety aspect of each gene construct, however, must be evaluated on the basis of its genetic background, specific gene product, and the environmental context of the host crop.

Resistance to disease based on a product from a single gene, like resistance to insects, is often overcome by development of new pest strains. Much concern has surrounded the development of plants engineered to produce their own pesticide and the subsequent development of resistant insects. Engineered resistance to insects has focused on the use of the gene for Cry (Bt) toxins, for controlling lepidopterous. Carefully management strategies will be further required to prevent development of resistance breaking pathogens, when disease-protected GM plants are grown commercially. Certainly, the breakdown of resistance as a result of pathogen adaptation, occurs in cultivars developed by classical breeding and can be anticipated if single genes are used for engineering disease-protected GM plants (Wally & Punja, 2010). Moreover, evolution of recombinant pathogens has been raised as a particular concern in strategies using structural viral genes that could recombine with naturally infecting viruses to form new viral forms (Regev et al., 2006).

Finally, the possibility of increased weediness of plants is particularly relevant to the design of plants with herbicide tolerance. Genetically modified plants themselves could become weedy and difficult to control on account of their resistance to a particular herbicide of equal concern is movement of the herbicide-tolerance trait to weedy relatives by pollen transfer. The invasiveness of any species and long-term environmental effects are difficult to project for any modified plant released into the ecosystem. However, it is not well established whether genetic modifications created by molecular techniques pose a greater risk of invasiveness than those created by classical breeding techniques (Jacobsen & Schouter, 2007).

2.2.1 Food safety and side effects of GM-crops

The safety of GM foods has been a controversial issue over the past decade. Despite major concerns, very little independent research has been carried out to establish their long-term safety.

In general terms, the safety assessment of GM foods should investigate: a) toxicity, b) allergenicity, c) specific components thought to have nutritional or toxic properties, d) stability of the inserted gene, e) nutritional effects associated with genetic modification, and f) any unintended effects which could result from the gene insertion (WHO, 2002).

Construction of transgenic crop plants must take into consideration any possible impact on food safety, concretely in the two areas of concern which are allergenicity and the production of gene products that are toxic to mammalian metabolism (Uzogara, 2000; Malarkey, 2003; Dona & Arvanitoyannis, 2009).

Biotechnology companies argue that many of the individual proteins used in GM crops have been consumed over a long period in their natural host with no health effects seen, so simply creating the proteins in a new plant will surely be the same. This assumes both that the new protein in the GM plant is identical to the naturally produced protein, and that no unintended effects have occurred during the genetic modification process that could produce other proteins. This assumption can be seen in action in the USA, where after ten years of commercialisation of GM crops there is still no post-market surveillance for allergic reactions (Davies, 2005). New research now challenges this assumption, and brings into question the safety of both new and previously approved GM foods.

About 1-2% of adults population in the world and about 5% of children display food allergies. Around 90% of food allergies are induced by peanuts, soybeans, vegetables, fruits, milk, eggs, cereals, nuts, some fish and shellfish. Generally speaking, the allergic reaction is caused not by whole food items, but only by certain components called allergens, which most commonly are proteins, or in fact only segments of proteins (peptides) called allergenic epitopes. Biotechnology allows crop breeders to add new genes to a plant, but also to remove or inactivate a specific gene. This opens the possibility of removing specific allergens so that those people who suffer from a specific food allergy can again eat that GM food. Such "allergen-free" foods have not yet come on the market, but they are being developed in various laboratories. One group in Japan reported several years ago that they had removed the major allergen from a variety of rice. In the US research is being done to remove the main allergen from peanuts and shrimps (Randhawa et al., 2011).

Allergenicity Many children in the US and Europe have developed life-threatening allergies to peanuts and other foods. There is a possibility that introducing a gene into a plant may create a new allergen or cause an allergic reaction in susceptible individuals.

A proposal to incorporate a gene from Brazil nuts into soybeans (a methionine-rich protein) was abandoned because of the fear of causing unexpected allergic reactions (Nordlee et al., 1996). Some people are allergic to proteins that occur naturally in soybeans, and they could have a reaction if they are exposed to either conventional or transgenic soybeans or soy products. Soybeans are one of the eight most common sources of food allergies. Although less common, some people have food allergies associated with corn and they could be affected by either conventional or transgenic corn. No allergic reactions attributable to the proteins present as a result of genetic engineering have been reported in the transgenic soybeans being grown commercially at this time. Reports of an allergenic protein made as a result of genetic engineering in one particular type of transgenic corn could not be confirmed by subsequent testing.

While there isn't any evidence that allergens have been introduced into food crops by genetic engineering, two incidents have received quite a bit of publicity and caused public concern about food allergies resulting from transgenic crops:

The first incident involved soybean plants, and a gene from Brazil nuts to make soybeans that contained higher levels of the amino acid methionine, to improve nutritious chicken feed that would eliminate the need for expensive feed supplements.

The second incident involved reports of allergic reactions in people who may have eaten food containing the insecticidal protein called Cry9C, one of several forms of the Bt insecticide. When food from grocery shelves tested positive for Cry9C, demonstrating an accidental way into the food supply. During this time, the reports surfaced of allergic reactions in people who had eaten corn products that may have been contaminated by Cry9C, but special test developed by the FDA (an enzyme-linked immunosorbent assay, or ELISA test, to detect people's antibodies to the Cry9C protein) did not find any evidence that the reactions in the affected people were associated with hypersensitivity to the Cry9C protein. The test isn't 100% conclusive, though, partly because food allergies may sometimes occur without detectable levels of antibodies to allergens. Extensive testing of GM foods may be required to avoid the possibility of harm to consumers with food allergies, and labeling of GM foods and food products will acquire new importance.

Before any GMO or derived product can be marketed in the EU, it must pass through an approval system which is intended to assess its safety for humans, animals and the environment. The GMO Panel of the European Food Safety Authority (EFSA), which provides scientific advice and technical support for GM food safety issues, published guidance for applicants seeking authorisation of GM food and/or feed, and a section of the guidance covers current requirements for assessment of allergenicity. The guidelines are based on the recommendations of the Codex Alimentarius Commission's ad hoc Intergovernmental Task Force on Foods Derived from Biotechnology. Codex is an organisation that develops international standards for food standards (Codex Procedural Manual 20th *Ed.* 2011, WHO/FAO).

Toxicity of transgene products in new and another major concern about GMOs, regardless of the source of the gene sequence. There is a growing concern that introducing foreign genes into food plants may have an unexpected and negative impact on human health. Various food plants produce compounds that would be toxic at high levels and enhancement of their production above normal levels in transgenic plants could be detrimental to human health. Up regulation of glycoalkaloids in potato by genetic manipulation, for example, would be of concern whether the increased levels resulted from classical breeding or genetic engineering (Friedman & McDonald, 1997). The gene introduced into the potatoes was a snowdrop flower lectin, a substance known to be toxic to mammals. The scientists who created this variety of potato chose to use the lectin gene simply to test the methodology, and these potatoes were never intended for human or animal consumption.

Side effects of gene products expressed in transgenic plants are also not to be ignored. The possible effect of Bt endotoxin in pollen ingested by monarch butterflies has received public attention, although further field research demonstrated low impact on lepidopterans that could be at high potential risk (Losey et al., 1999). In additinon, concerns about secondary effects on non-target insects must be balanced by the impact of traditional pesticides that would have been used if Bt transgenics were not grown.

Other example of possible secondary effects can be envisioned, such as the effect of antifungal or antibacterial gene products on degradation of crop residues in the field. Decrease in plant decomposition could affect soil fertility and, in some cases, antimicrobial gene products could lower the diversity of soil microorganism communities (Wolfenbarger & Phifer, 2000).

Whether or not these effects are important for the environment or ecology of particular macro- or micro-organisms is something that needs to be evaluated within the scope of establishing the safety of any particular transgenic plant species.

2.2.2 Controversy about the safety to eat GM crops

The primary concern many people have about genetically engineered (GE) crops is the safety of food made from them. It is unlikely that eating DNA poses any significant risk to human or animal health, and there is no evidence to suggest that there is any additional risk from the transgenes present in genetically engineered plants. Although there continues to be quite a bit of controversy over this issue, no evidence has been found that foods made with the genetically engineered crops now on the market are any less safe to eat than foods made with the same kinds of conventional crops.

The overall goal about GE crops is not to establish an absolute level of safety, but rather the relative safety of the new product so that there is a reasonable certainty that no harm will result from intended uses under the anticipated conditions of production, processing and consumption. Most of the DNA we eat is degraded in the digestive system, but some experiments have shown that small amounts of it can be found in some cells in the body. It is thought to be unlikely that this DNA would be incorporated into the DNA of those cells, but even if it was, the chance of any undesirable effect on the whole organism is thought to be very low. Normal diets for humans and other animals contain large amounts of DNA. This DNA comes not only from the cells of the various kinds of plants or animals constituting the food, but also from any contaminating microorganisms or viruses that may be present in or on the food. We have been exposed to this variety of DNA throughout our entire history. It seems that we are well adapted to handling exposure to DNA, and there is no obvious reason that the DNA from other organisms introduced into crops by genetic engineering would have any additional effect.

Some critics of GE crops point out that a lack of evidence for harmful effects does not mean they do not exist, but just as likely could mean that we have not done the proper studies to document them. Some reject the idea that we face the same kinds of risks from GE crops as from conventionally developed crops, believing the genetic engineering process itself introduces unique risks. Genetically engineered crop varieties are being subjected to far greater scientific scrutiny than that ordinarily given to conventional varieties, even though many scientists have argued that there is no strict distinction between the food safety risks posed by genetically engineered plants and those developed using conventional breeding practices.

Safety assessments of foods developed using genetic engineering include the following considerations:

1. Evaluation of the methods used to develop the crop, including the molecular biological data which characterizes the genetic change,
2. The evaluation for the expected phenotype,
3. The general chemical composition of the novel food compared to conventional counterparts,
4. The nutritional content compared to conventional counterparts,
5. The potential for introducing new toxins, and
6. The potential for causing allergic reactions.

A major concern often expressed about GE food safety is the risk for unintentional, potentially harmful changes that may escape detection in the evaluation process. It is true that the number of factors that are examined for change is small compared to the total number of components produced by plants. Also, more extensive comparisons of plant chemical compositions would be difficult because complete data describing the composition of conventional crop plants, including knowledge of variability among different cultivars or that due to environmental influences, is lacking. The random nature of transgene insertion when making GE plants, it is argued, may cause disruption of important genes, causing significant effects but little obvious change to the plant's phenotype.

Antibiotic resistance genes are frequently used at several stages in the creation of GE plants as convenient "selectable markers". Bacteria or plant cells without a gene for resistance to the antibiotics used can be killed when the antibiotic is applied to them. So when scientists link the gene for the desired trait being introduced into a plant with an antibiotic resistance gene, they can separate cells carrying the desired gene from those that don't by exposing them to the antibiotic. The antibiotic resistance genes end up in the genetically engineered plants as excess baggage whose function is no longer required after the process of making them is complete. Concern has been raised about the possibility that antibiotic resistance genes used to make transgenic plants could be transferred to microorganisms that inhabit the digestive tracts of humans or other animals that eat them, and therefore might contribute to the already serious problem of antibiotic resistant pathogens. Transfer of DNA from one microbe to another (horizontal gene transfer) is known to occur in nature and has been observed in some laboratory experiments under specific conditions, but the likelihood of DNA being transferred from plant material in the digestive system to microbes has not yet been experimentally determined. It is thought that for such a transfer to be possible, it would have to come from consumption of fresh food since most processing would degrade the plant's DNA. Also, there is evidence that most DNA is rapidly degraded by the digestive system. Overall, the risk of antibiotic resistance genes from transgenic plants ending up in microorganisms appears to be low.

A second concern about the use of some antibiotic resistance genes is that they could reduce the effectiveness of antibiotics taken at the same time transgenic food carrying the resistance gene for that antibiotic was consumed. In cases where this has been identified as a risk based on the mechanism of resistance, studies have suggested the chance of this happening was probably very low due to rapid digestion of the inactivating enzymes produced by the transgenic resistance gene. Most transgenic plants do not carry resistance genes for antibiotics commonly used to treat infections in humans. Scientists are developing and using different selectable markers, and are also experimenting with methods for removing the antibiotic resistance genes before the plants are released for commercial use.

2.2.3 Advantages of transgenic plants

Despite the many concerns transgenic plants raise, they do have immense potential for benefit to society (Peterson et al., 2000).

Positive effects may include soil conservation, as new cultural practices permit low till methods and consequential maintenance of soil structure and decreased erosion.

Transgenic plants with stable resistance to disease will restrict crop losses and permit increased yield. Losses of food during postharvest storage can be decreased. There are also direct benefices to decreased use of pesticides and savings in resources and energy to manufacture and apply chemicals. Genetically modified plants may one day allow us to grow profitable crops without the need for environmentally unfriendly disease control plans. Globalization of the agricultural industry inevitably results in globalization of plant diseases. Various diseases, such as blight on potato, appear to be spreading worldwide, karnal bunt on wheat is on the increase in Asia and parts of North America, mosaic is increasing on cassava in Africa, and leaf blight continues to spread on rice in Japan and India (Moffat, 2001). Disease protected transgenic plants may yet demonstrate to be an important alternative against plant pathogens. Molecular biology has the potential to contribute significantly to a better society in which the environment is respected and an adequate food supply is provided.

Furthermore, the world population has topped 6 billion people and is predicted to double in the next 50 years. Ensuring an adequate food supply for this booming population is going to be a major challenge in the years to come. GM foods promise to meet this need in a number of ways: 1) Pest resistance Crop losses from insect pests can be staggering, resulting in devastating financial loss for farmers and starvation in developing countries, beside health hazards risks of chemical treatments, and contaminations of water and the environment; 2) Herbicide tolerance, avoiding utilization of chemicals to kill weed. Crop plants genetically-engineered to be resistant to one very powerful herbicide could help prevent environmental damage by reducing the amount of herbicides needed; 3) Disease resistance caused by viruses, fungi and bacteria; 4) Cold tolerance, avoiding destruction of sensitive seedlings; 5) Drought tolerance/salinity tolerance; 6) Nutrition. Some crops do not contain necessary nutrients to prevent malnutrition. GM crops are directed toward increase minerals and vitamine, i.e. β-carotene; 7) Pharmaceuticals production, such as edible vaccines in tomatoes and potatoes, which will be much easier to ship, store and administer than traditional injectable vaccines; 8) Phytoremediation. Soil and groundwater pollution continues to be a problem in all parts of the world. Plants have been genetically engineered to clean up heavy metal pollution from contaminated soil.

3. Unification, development and implementation of official standard technologies

The coexistence of GM plants with conventional and organic crops has raised significant concern in many European countries. Due to relatively high demand from European consumers for the freedom of choice between GM and non-GM foods, EU regulations require measures to avoid mixing of foods and feed produced from GM crops and conventional or organic crops. European research programs are investigating appropriate methods and tools to keep both GM and non GM crops isolated, i.e. isolation distance and pollen barriers, which are usually not used in North America because they are very costly and there are no safety-related reasons to employ them (Ramessar et al., 2010).

Certain global regulatory bodies require development of DNA detection methods that allow for unique identification of commercial transgenics, harmonised guidelines for the validation and use of these methods are not yet in place. As a result, numerous governmental agencies, global standards organizations, and industry organisations are attempting to develop their own independent standardisation guidelines for testing methodologies.

In the European Union, the (JRC) is playing a leading role in ensuring a harmonised approach between EU Member States, industry and stakeholders. It now hosts six European Union Reference Laboratories (EU-RLs) on food and feed safety in support of EU Member States' National Reference Laboratories (NRLs) in the respective fields. It is the National

Food Authorities who are responsible for the appropriate implementation of legislation. The latter is in place both to ensure the safety and quality of food products including animal feed and to ensure public health.

In order to ensure public health, potentially hazardous residues and contaminants are put under vigorous scrutiny and strict authorisation procedures for new additives and crops for feed and food production are in place. The aim of EU-RLs is to guarantee uniform detection, quantification and authorisation procedures. The activities of EU-RLs cover all the areas of feed and food law and animal health. In particular, those areas where there is a need for precise analytical and diagnostic results. The main objective of the EU-RLs is to contribute to a high quality and uniformity of results obtained in the various official food and feed control laboratories throughout the European Union.

Two JRC EU-RLs support authorisation for additives for feed production and of crops to be used in food and feed that have been genetically modified i.e. containing genetically modified organisms (GMOs). This work is carried out in close collaboration with the European Food Safety Authority (EFSA), the latter being responsible for risk assessment of such new substances and crops.

The main responsibilities of the EU-RLs for feed and food are to: 1) provide National Reference Laboratories (NRLs) with details of analytical methods, including reference methods, 2) organise comparative (proficiency) testing amongst the NRLs, 3) conduct training courses for the benefit of staff from the NRLs and of experts from developing countries, and 4) provide scientific and technical assistance to the European Commission, especially in cases when Member States contest the results of analyses.

The work of the EU-RLs contributes to increasing European and worldwide standardisation of analytical methods. This helps to ensure that the quality of analytical data obtained in various laboratories are increasingly comparable. Methods are developed by EU-RLs and then validated through collaborative trial testing in collaboration with the NRLs and other expert laboratories in the respective field. Proficiency tests are also organised by the EU-RLs (for NRLs) and by the NRLs (for national official laboratories) to ensure the quality of data obtained in the various laboratories that are also required for European and other international monitoring databases for exposure and risk assessment.

In this way, EU-RLs are working towards the best interests of the consumer. They are helping to build confidence in the results obtained by food control laboratories and to ensure that products purchased are in compliance with legislation and have the highest food hygiene standards.

EU-RLs also represent a unique platform for information exchange on analytical methodology and quality assurance tools for control laboratories. Together with the network of NRLs, they provide a pool of knowledge and facilities that makes them best placed to handle emerging issues.

The JRC is currently managing six EU-RLs. These are located in the Institute for Reference Materials and Measurements (IRMM) in Belgium and the Institute for Health and Consumer Protection (IHCP) in Italy. It is worth noting that the JRC-IRMM also chairs a board of expert laboratories which acts as EU-RL on behalf of the European Commission's Directorate General Agriculture. Its purpose is to harmonise analytical methodologies for the determination of water content in poultry to ensure the quality and to prevent fraud: 1) EU-RL for GMOs in food and feed, 2) EU-RL for feed additives, 3) EU-RL for food contact materials, 4) EU-RL for heavy metals in feed and food, 5) EU-RL for mycotoxins in food and feed, and 6) EU-RL for polycyclic aromatic hydrocarbons.

Regulation of genetically engineered crops by US government is made because guidance of the first the federal government adopted a "Coordinated Framework for Regulation of Biotechnology". Under this system, three federal agencies have regulatory authority over genetically engineered (GE) crops. Each agency has a different role to ensure safety under specific legislation. These agencies and their regulatory responsibilities are:

1) *The U.S. Department of Agriculture* (USDA), through the Animal and Plant Health Inspection Service (APHIS), is responsible for assuring that any organism, including genetically engineered organisms, will not become pests that can cause harm if they are released into the environment. APHIS has used their authority to grant permission and set the rules for field testing of genetically engineered crops. These crops cannot be commercialized until they are granted "non regulated" status by APHIS upon satisfactory review of the field testing data.

2) *The Food and Drug Administration* (FDA) is responsible for ensuring the safety of most food (except for meat, poultry and some egg products, which are regulated by the U.S. Department of Agriculture), including food from genetically engineered crops. If the allergen, nutrient and toxin content of new GE foods fall within the normal range found in the same kind of conventional food, the FDA does not regulate the GE food any differently. So far, all genetically modified foods in the U.S. marketplace have gone through a voluntary review process where the FDA determines whether they are "not substantially different" from the same conventional foods by consulting with developers of new GE foods to identify potential sources of differences, then reviewing a formal summary of data provided by the developer. Recently, the FDA has announced a new rule that would make pre-market consultation mandatory. The FDA has the authority to order foods to be pulled from the market at any time if are found to be unsafe, or to require labeling of any food that has different amounts of allergens, nutrients, or toxins than a consumer would expect to find in that kind of food.

3) *The Environmental Protection Agency* (EPA) evaluates the safety of any pesticides that are produced by genetically engineered plants. The EPA calls novel DNA and proteins genetically engineered into plants to protect them against pests "plant incorporated protectants" (PIPs) and regulates them the same way they regulate other pesticides.

Under the Coordinated Framework, some kinds of genetically engineered crops might not be subject to the oversight of all three agencies. For example, an ornamental flower like petunias engineered to have longer lasting blooms may only have to meet the requirements of APHIS, but a food crop like soybeans engineered to produce an insecticidal compound would be subject to the rules of all three agencies. Additional regulations are imposed by some states. Also, the National Institutes of Health has developed safety procedures for research with recombinant DNA. Most institutions developing genetically engineered crops follow the NIH guidelines, and they are required for federally funded research.

4. Labeling of genetically engineered foods

GMO labelling was introduced to give consumers the freedom to choose between GMOs and conventional products. Essentially, if a foodstuff is produced using genetic engineering, this must be indicated on its label. The target of most labeling efforts is food products that were genetically engineered, that is, they contain genes artificially inserted from another organism.

Whether or not to require mandatory labeling of genetically engineered (GE) foods is a major issue in the debate over the risks and benefits of food crops produced using biotechnology. The issue is complex because 1) many arguments put forth in the debate are based on disagreements about the adequacy of our scientific understanding of the consequences of genetic engineering; and (2) significant changes to our current food marketing and manufacturing system, with potentially large economic impacts, would be required to implement mandatory labeling.

Actual labelling practice, however, is far more complicated, and must be planned and regulated with issues such as feasibility, legal responsibilities, coherence and standardisation in mind.

While some groups advocate the complete prohibition of GMOs, others call for mandatory labeling of genetically modified food or other products. Other controversies include the definition of patent and property pertaining to products of genetic engineering. According to the documentary Food, Inc. efforts to introduce labeling of GMOs has repeatedly met resistance from lobbyists and politicians affiliated with companies developing GM crops.

Governments around the world are hard at work to establish a regulatory process to monitor the effects of and approve new varieties of GM plants. Yet depending on the political, social and economic climate within a region or country, different governments are responding in different ways.

Agribusiness industries believe that labeling should be voluntary and influenced by the demands of the free market. If consumers show preference for labeled foods over non-labeled foods, then industry will have the incentive to regulate itself or risk alienating the customer. Consumer interest groups, on the other hand, are demanding mandatory labeling.

Central to the arguments for mandatory labeling is that consumers have the right to know what they are eating. This is especially true for some products made with biotechnology where health and environmental concerns have not been satisfactorily resolved. Historically industry has proven itself to be unreliable at self-compliance with existing safety regulations. Some people do not wish to use genetically engineered products for religious or ethical reasons. Labeling is the only way consumers can make informed choices, whatever their reasons may be.

Major arguments against mandatory labeling have addressed the practical concerns about the expense and complex logistics that would be required to ensure GE and conventional foods are kept separate or to test all foods for GE content. It is argued that such measures are unnecessary since no significant differences have been found between today's GE foods and conventional foods. Enacting mandatory labeling will also require resolving certain other questions. Major issues include defining exactly what kinds of technologies would be covered, deciding on tolerance levels for genetically engineered content or ingredients before labeling would be required, and choosing a method for verifying that products are properly labeled.

In Europe, anti-GM food protestors have been especially active. In the last few years Europe has experienced two major foods scares: bovine spongiform encephalopathy (mad cow disease) in Great Britain and dioxin-tainted foods originating from Belgium. These food scares have undermined consumer confidence about the European food supply, and citizens are disinclined to trust government information about GM foods, establishing a mandatory food labeling of GM foods in stores, with a 1% threshold for contamination of unmodified foods with GM food products, a commonly proposed threshold. In other words, if any ingredient of a product exceeds one percent GM content, the product needs labeling. One percent is the labeling threshold decided upon by Australia and New Zealand. The European Union has decided on a level of 0.9 percent, while Japan has specified a five percent threshold. Thresholds as low as 0.01 percent (the approximate limit of detection) have been recommended (Davison, 2010).

The shift of global agriculture towards biotech varieties, however, has not been supported by all elements of society. In response to these differing levels of acceptance of the use of this technology, several countries have adopted regulations requiring that foods prepared from GM ingredients be labeled as such. However, labeling of foods is necessary only when the concentration of GM material in a food ingredient measures above a specified threshold concentration (%GM). The adoption and implementation of such laws can have significant consequences to global commerce in agriculture, food, and feed. Meeting these global market requirements for GM compliances is further complicated by the fact that each country has different regulations, including different GM ingredient thresholds for labeling and different methods of testing.

In the United States, the regulatory process is confused because there are three different government agencies that have jurisdiction over GM foods. The EPA evaluates GM plants for environmental safety, the USDA evaluates whether the plant is safe to grow, and the FDA evaluates whether the plant is safe to eat. The EPA is responsible for regulating substances such as pesticides or toxins that may cause harm to the environment. GM crops such as B.t. pesticide-laced corn or herbicide-tolerant crops but not foods modified for their nutritional value fall under the purview of the EPA. The USDA is responsible for GM crops that do not fall under the umbrella of the EPA such as drought-tolerant or disease-tolerant crops, crops grown for animal feeds, or whole fruits, vegetables and grains for human consumption. The FDA historically has been concerned with pharmaceuticals, cosmetics and food products and additives, not whole foods. Under current guidelines, a genetically-modified ear of corn sold at a produce stand is not regulated by the FDA because it is a whole food, but a box of cornflakes is regulated because it is a food product. The FDA's stance is that GM foods are substantially equivalent to unmodified, "natural" foods, and therefore not subject to FDA regulation.

Independently of the government organism, there should be regulated the verification claims to know if a food is or is not genetically engineered. There are two ways this can be done: 1) Content-based verification requires testing foods for the physical presence of foreign DNA or protein. A current application of this type of procedure is the analysis and labeling of vitamin content of foods. As the number of transgenes in commercialized crops increases, the techniques for detecting an array of different transgenes have become more sophisticated (Shrestha et al., 2008). 2) Process-based verification entails detailed record-keeping of seed source, field location, harvest, transport, and storage (Sundstrom et al., 2002).

There are many questions that must be answered whether labeling of GM foods becomes mandatory:

First, it is concern about whether consumers are willing to absorb the cost of labeling. Accurate labeling requires an extensive identity preservation system from farmer to elevator to grain processor to food manufacturer to retailer (Maltsbarger & Kalaitzandonakes, 2000). If the food production industry is required to label GM foods, factories will need to construct two separate processing streams and monitor the production lines accordingly. Farmers must be able to keep GM crops and non GM crops from mixing during planting, harvesting and shipping. It is almost assured that industry will pass along these additional costs to consumers in the form of higher prices. Either testing or detailed recordkeeping needs to be done at various steps along the food supply chain. Estimates of the costs of mandatory labeling vary from a few dollars per person per year to 10 percent of a consumer's food bill (Gruere & Rao, 2007). Consumer willingness to pay for GE labeling information varies widely according to a number of surveys, but it is generally low in North America. Another potential economic impact for certain food manufacturers is that some consumers may avoid foods labelled as containing GE ingredients.

Secondly, the acceptable limits of GM contamination in non GM products. The EC has determined that 1% is an acceptable limit of cross-contamination, yet many consumer interest groups argue that only 0% is acceptable. In addition, it is necessary to know who is going to monitor these companies for compliance and what could be the penalty if they fail.

Third, concerns the level of traceability of GM food cross-contamination. Scientists agree that current technology is unable to detect minute quantities of contamination, so ensuring 0% contamination using existing methodologies is not guaranteed. Yet researchers disagree on what level of contamination really is detectable, especially in highly processed food products such as vegetable oils or breakfast cereals where the vegetables used to make these products have been pooled from many different sources. A 1% threshold may already be below current levels of traceability.

Finally, who should be responsible for educating the public about GM food labels. Food labels must be designed to contain clearly and accurate information about the product in simple language that everyone can understand. This may be the greatest challenge faced be a new food labeling policy: how to educate and inform the public without damaging the public trust and causing alarm or fear of GM food products.

Under current policy, the U.S. Food and Drug Administration do not automatically require all genetically engineered food to be labeled. Conventional and genetically engineered (GE) foods are all subject to the same labeling requirements, and both may require special labeling if particular food products have some property that is significantly different than what consumers might reasonably expect to find in that kind of food. Therefore, particular genetically engineered foods are subject to special labeling requirements if the FDA concludes they have significantly different properties including:

1. A different nutritional property from the same kind of conventional food,
2. A new allergen consumers would not expect to be in that kind of food (a hypothetical example would be an allergenic peanut protein in GE corn or some other crop),
3. a toxicant in excess of acceptable limits.

Examples of genetically engineered foods that require special labeling are those that contain vegetable oil made from varieties of GE soybeans and canola where the fatty acid composition of the oils extracted from the seeds of these crops was altered. Since the oils from these varieties have different nutritional properties than conventional soy and canola oils, foods made with them must be labeled to clearly indicate how they are different. You might see "high laurate canola" or "high oleic soybean" on food labels if these products were used. The FDA does not require them to be labeled as "genetically engineered", but that information could also be included on the label.

So far, no approved, commercially grown genetically engineered food crops have known properties that would require foods made from them to be labeled because they contain a new allergen or excess levels of toxic substances.

Federal legislation has been proposed that would require mandatory labeling of genetically engineered foods and similar initiatives at the state or local level have been considered or are currently pending.

5. Conclusions

The present atmosphere surrounding genetically engineered crops has led to a situation where food safety assessment is not just about science, but also about concerns, and standards about how to assure "safety." The detection, identification and quantification of the GMO content in food or feed products are a great challenge. The existing analytical methods for GMO testing leave the inspection authorities with many choices and compromises.

Methods, which can guarantee absence of non approved GMOs in seed samples even at the suggested 0.1% level of GM contamination does not exist at present. However, PCR and immunoassay based technologies are often used for the detection of products of agricultural biotechnology. They are valuable and reliable tools for the detection of GM products in seed production and very early in the food and feed supply chain. Concretely, when operated within specifications, immunoassays have been proven, in most cases, to be fast, reliable, and economic test methods.

It is critical that such methods are reliable and give the consistent results in laboratories across the world. This includes the need for a proper validation of the methods. The choice of the appropriate reference material will impact the reliability and accuracy of the analytical results, and numerous biological and analytical factors need to be taken into account when reporting results. Furthermore, as scientific opportunities advance, agreement on reasonable standards of safety for developing countries will be critical, and exchange of data as well, which will help ensure that data requirements are manageable across the developing world.

We don't know yet all the potential risks that the GMO could have, by long term accumulation, upon the environment. New strategies, therefore, are required to face the continuing challenge of disease spread to new environments and emergence of resistance-breaking strains of microbial plant pathogens. Disease-protected transgenic plants may yet prove to be an important arsenal in the battle against plant pathogens. With a judicious approach and careful development of new innovations, molecular biology has the potential to contribute significantly to a better society in which the environment is respected and an adequate food supply is provided. Genetically-modified foods have the potential to solve many of the world's hunger and malnutrition problems, and to help protect and preserve the environment by increasing yield and reducing reliance upon chemical pesticides and herbicides. Yet there are many challenges ahead for governments, especially in the areas of safety testing, regulation, international policy and food labeling.

The achievements of the genetic engineering have nowadays considerable benefits, but now we don't know the price we, or the future generations, will have to pay for these benefits. The long term risks of the GMO are not entirely known today, and long-term studies are clearly necessary. We must proceed with caution to avoid causing unintended harm to human health and the environment as a result of our enthusiasm for this powerful technology.

Globalization of the agricultural industry inevitably results in globalization of markets, so competency in assuring food safety for GM crops is essential. This competency will enable countries to conduct independent research when necessary. Building such capacity also creates sufficient infrastructure to allow scientifically defensible decisions in the face of food safety questions colored by each country's perceptions and circumstances. It is obvious that international collaboration is needed to ensure that the methods offered by the different companies hold promises, which can be done by elaboration of:

1. Further research to understand the appropriateness of DNA and protein based methodologies,
2. Compatibility between methods,
3. Appropriate protocols for validation studies and for proficiency testing,
4. Make appropriate reference materials readily and globally available.

In the future, the number of different GMOs is expected to grow, and research is going in the direction to develop GM plants with inducible promoters that activate specific traits when needed. It can be expected that detection of GMOs will become more complicated in the near future, being one of the mayor challenges for the future will be to develop analytical identification methods that facilitate screening for all the promoters used worldwide.

6. Acknowledgment

MCH-S thanks the Fulbright program and the Spanish Ministry of Education for a postdoctoral fellowship (FMECD-2010).

7. References

Ahmed, F.E. (2002). Detection of genetically modified organisms in foods. *Trends Biotechnol.*, Vol.20, pp.215–223

Bhatnagar-Mathur, P., Vadez, V. & Sharma, K.K. (2008). Transgenic approaches for abiotic stress tolerance in plants: retrospect and prospects. *Plant Cell Rep.*, Vol.27, No.3, pp.411-424

Brown, M.E. & Funk, CC. (2008). Climate. Food security under climate change. *Science*, Vol.319, pp.580–581

Codex Procedural Manual 20th Ed. WHO/FAO and Consideration of the Methods for the Detection and Identification of Foods Derived from Biotechnology; General Approach and Criteria for Methods (2011) Codex Alimentarius, Rome, Italy. ftp://ftp.fao.org/codex/Publications/ProcManuals/Manual_20e.pdf

Davies, H.V. (2005). GM organisms and the EU regulatory environment: allergenicity as a risk component. *Proc Nutr Soc.*, Vol.64, No.4, pp.481-486

Davison, J. (2010). GM plants: Science, politics and EC regulations. *Plant Science*, Vol.178, No.2, pp.94–98

De Boer, S.H. (2003). Perspective on genetic engineering of agricultural crops for resistance to disease. *Can. J. Plant Pathol.*, Vol25, pp.10-20

Dona, A. & Arvanitoyannis, I.S. (2009). Health risks of genetically modified foods. *Crit Rev Food Sci Nutr.*, Vol.49, No.2, pp.164-75

Dörries, H.H., Remus, I., Grönewald, A., Grönewald, C. & Berghof-Jäger, K. (2010). Development of a qualitative, multiplex real-time PCR kit for screening of genetically modified organisms (GMOs). *Anal Bioanal Chem.*, Vol.396, No.6, pp.2043-2054

Fantozzi, A., Ermolli, M., Marini, M., Scotti, D., Balla, B., Querci, M., Langrell, S.R. & Van den Eede, G. (2007). First application of a microsphere-based immunoassay to the detection of genetically modified organisms (GMOs): quantification of Cry1Ab protein in genetically modified maize. *J Agric Food Chem.*, Vol.55, No.4, pp.1071-1076

Friedman, M. & McDonald, G.M. (1997). Potato glycoalkaloids: chemistry, analysis, safety and plant physiology. *Crit. Rev. Plant Sci.*, Vol.16, pp.55–132

Grothaus, G.D., Bandla, M., Currier, T., Giroux, R., Jenkins, G.R., Lipp, M., Shan, G., Stave, J.W., Pantella, V. (2006). Immunoassay as an analytical tool in agricultural biotechnology. *J AOAC Int.*, Vol.89, No.4, pp.913-28

Gruere, G.P. & Rao, S.R. (2007). A review of international labeling policies of genetically modified food to evaluate India's propose rule. *AgBioForum*, Vol.10, No.1

Hüsken, A. & Dietz-Pfeilstetter, A. (2007). Pollen-mediated intraspecific gene flow from herbicide resistant oilseed rape (*Brassica napus* L.). *Transgenic Res.*, Vol.16, No.5, pp.557-569

Ingham, D.J., Beer, S., Money, S., & Hansen, G. (2001). Quantitative Real-Time PCR Assay for Determining Transgene Copy Number in Transformed Plants. *Biotechniques*, Vol.31, pp.132–134

Jacobsen, E. & Schouten, H.J. (2007). Cisgenesis strongly improves introgression breeding and induced translocation breeding of plants. *Trends Biotechnol.*, Vol.25, No.5, pp.219-23

Jimenez-Lopez, J.C., Gachomo, E.W., Ariyo, O.A., Baba-Moussa, L. & Kotchoni, S.O. 2012. Specific conformational epitope features of pathogenesis-related proteins mediating cross-reactivity between pollen and food allergens. *Mol. Biol. Rep.* 39(1): 123-130

Lipton, C.R., Dautlick, J.X., Grothaus, C.D., Hunst, P.L., Magin, K.M., Mihaliak, C.A., Rubio, F.M., & Stave, J.W. (2000). Guidelines for the validation and use of immunoassays for determination of introduced proteins and biotechnology enhanced crops and derived food ingredients. *Food Agric. Immunol.*, Vol.12, pp.153–164

López, M.C., Garcia-Cañas, V. & Alegre, M.L. (2009). Reversed-phase high-performance liquid chromatography-electrospray mass spectrometry profiling of transgenic and non-transgenic maize for cultivar characterization. *J Chromatogr A.*, Vol.1216, No.43, pp.7222-7228

Losey, J.E., Rayor, L.S. & Carter, M.E. (1999). Transgenic pollen harms monarch butterfly. *Nature*, Vol.399, pp.214

Malarkey, T. (2003). Human health concerns with GM crops. *Mutat Res.*, Vol.544, No.2-3, pp.217-221

Maltsbarger, R. & Kalaitzandonakes, N. (2000). Direct and hidden costs in identity preserved supply chains. *AgBio Forum*, Vol.3, No.4

Matsuoka, T., Kawashima, Y., Akiyama, H., Miura, H., Goda, Y., Kusakabe, Y., Isshiki, K., Toyoda, M., & Hino, A. (2000). A method of detecting recombinant DNAs from four lines of genetically modified maize. *J. Food Hyg. Soc. Jpn.*, Vol.41, pp.137–143

Michelini, E., Simoni, P., Cevenini, L., Mezzanotte, L. & Roda, A. (2008). New trends in bioanalytical tools for the detection of genetically modified organisms: an update. *Anal Bioanal Chem*, Vol.392, No.3, pp.355-67

Moffat, A.S. (2001). Finding new ways to fight plant disease. *Science*, Vol.292, pp.2270–2273

Nasiruddin, K.M. & Nasim, A. (2007). Development of agribiotechnology and biosafety regulations used to assess safety of genetically modified crops in Bangladesh. *J AOAC Int.*, Vol.90, No.5, pp.1508-1512

Niemann, H. & Kues, W.A. (2007). Transgenic farm animals: an update. *Reprod Fertil Dev.*, Vol.19, No.6, pp.762-70

Nordlee, J.A., Taylor, S.L., Townsend, J.A., Thomas, L.A. & Bush, R.K. (1996). Identification of a Brazil-nut allergen in transgenic soybeans. *N. Engl. J. Med.*, Vol.334, No.688–694

Peterson, G., Cunningham, S., Deutsch, L., Erickson, J., Quinlan, A., Raez-Luna, E., Tinch, R., Troell, M., Woodbury, P., & Zens, S. (2000). The risks and benefits of genetically modified crops: a multidisciplinary perspective. Conservation Ecology, Vol.4, No.1, pp.13

Pollegioni, L., Schonbrunn, E. & Siehl, D. (2011). Molecular basis of glyphosate resistance-different approaches through protein engineering. *FEBS J.*, Vol.278, No.16, pp.2753-2766

Querci, M., Van den Bulcke, M., Zel, J., Van den Eede, G. & Broll, H. (2010). New approaches in GMO detection. *Anal Bioanal Chem.*, Vol.396, No.6, pp.1991-2002

Ramessar, K., Capell, T., Twyman, R.M. & Christou, P. (2010). Going to ridiculous lengths—European coexistence regulations for GM crops. *Nature Biotechnol.*, Vol.28, pp.133–136

Randhawa, G.J. & Singh, M. & Grover, M. (2011). Bioinformatic analysis for allergenicity assessment of *Bacillus thuringiensis* Cry proteins expressed in insect-resistant food crops. *Food Chem. Toxicol.*, Vol.49, No.2, pp.356-362

Regev A, Rivkin H, Gurevitz M, Chejanovsky N. (2006). New measures of insecticidal efficacy and safety obtained with the 39K promoter of a recombinant baculovirus. *FEBS Lett.*, Vol.580, No.30, pp.6777-6782

Sayanova, O. & Napier, J.A. (20110. Transgenic oilseed crops as an alternative to fish oils. *Prostaglandins Leukot Essent Fatty Acids*, doi:10.1016/j.plefa.2011.04.013

Schmidt, M.A. & Parrott, W.A. (2001). Quantitative detection of transgenes in soybean [Glycine max (L.) Merrill] and peanut (*Arachis hypogaea* L.) by real-time polymerase chain reaction. *Plant Cell Rep.*, Vol.20, pp.422–428

Schmidt, A.M., Sahota, R., Pope, D.S., Lawrence, T.S., Belton, M.P. & Rott, M.E. (2008). Detection of genetically modified canola using multiplex PCR coupled with oligonucleotide microarray hybridization. *J Agric Food Chem.*, Vol.56, No.16, pp.6791-800

Shrestha, H.K., Hwu, K.K., Wang, S.J., Liu, L.F. & Chang, M.C. (2008). Simultaneous detection of eight genetically modified maize lines using a combination of event- and construct-specific multiplex-PCR technique. *J Agr. Food Chem.*, Vol.56, pp.8962–8968

Song, P., Cai, C.Q., Skokut, M., Kosegi, B.D. & Petolino, J.F. (2002). Quantitative real-time PCR as a screening tool for estimating transgene copy number in WHISKERS™-derived transgenic maize. *Plant Cell Rep.*, Vol.20, pp.948–954

Sundstrom, F.J., Williams, J., VanDeynze, A. & Bradford, K.J. (2002). Identity preservation of agricultural commodities. *Agricultural Biotechnology in California Series*, Pub.8077

Takeda, S. & Matsuoka, M. (2008). Genetic approaches to crop improvement: responding to environmental and population changes. *Nat. Rev. Genet.*, Vol.9, pp.444–457

Trapmann, S., Schimmel, H., Kramer, G.N., Van den Eede, G. & Pauwels, J. (2002). Production of certified reference materials for the detection of genetically modified organisms. *J AOAC Int.*, Vol.85, No.3, pp.775-779

Trapmann, S., Corbisier, P., Schimmel, H. & Emons, H. (2010). Towards future reference systems for GM analysis. *Anal Bioanal Chem.*, Vol.396, No.6, pp.1969-75

Turner, W.R., Oppenheimer, M., Wilcove, D.S. (2009). A force to fight global warming. *Nature*, Vol.462, pp.278–279

Uzogara, S.G. (2000). The impact of genetic modification of human foods in the 21st century: a review. *Biotechnol Adv.*, Vol.18, No.3, pp.179-206

Wally, O. & Punja, Z.K. (2010). Genetic engineering for increasing fungal and bacterial disease resistance in crop plants. *GM Crop*, Vol.1, No.4, pp.199-206

Warwick, S.I., Beckie, H.J. & Hall, L.M. (2009). Gene flow, invasiveness, and ecological impact of genetically modified crops. *Ann N Y Acad Sci.*, Vol.1168, pp.72-99

Windels, P., Taverniers, I., Depicker, A., Van Bockstaele, E., & De Loose, M. (2001). Characterization of the Roundup Ready soybean insert. *Eur. Food Res. Technol.*, Vol.213, pp.107–112

Wolfenbarger, L.L. & Phifer, P.R. (2000). Ecological risks and benefits of genetically engineered plants. *Science*, Vol.290, pp.2088–2093

Zhu, M., Yu M. & Zhao S. (2009). Understanding quantitative genetics in the systems biology era. *Int J Biol Sci.*, Vol.5, No.2, pp.161-170

Part 3

Protection and Health Benefits

Control of *Salmonella* in Poultry Through Vaccination and Prophylactic Antibody Treatment

Anthony Pavic[1,2,4], Peter J. Groves[2,3] and Julian M. Cox[4]
[1]*Birling Avian Laboratories;*
[2]*School of Veterinary Science, University of Sydney;*
[3]*Zootechny;*
[4]*School of Biotechnology and Biomolecular Sciences,*
Faculty of Science, University of New South Wales
Australia

1. Introduction

Salmonella are ubiquitous, host-adapted or zoonotic human and animal pathogens and, after *Campylobacter*, the genus is the second most predominant bacterial cause of foodborne gastroenteritis Worldwide (Giannella, 1996; Smith, 2003). Numerous foods (meat, seafood, eggs, fresh horticulture, table eggs and poultry) have been associated with *Salmonella* carriage or contamination (FAO/WHO, 2002; Jay *et al.*, 2003; Foley and Lynn, 2008).

Poultry meat has been associated frequently and consistently with the transmission to humans of enteric pathogens, including *Salmonella* and *Campylobacter* (Food and Agriculture Organisation of the United Nations and the World Health Organisation [FAO/WHO], 2002; Codex Alimentarius Commission, 2004; EFSA, 2007; Lutful Kabir, 2010; Vanderplas *et al.*, 2010 and Cox and Pavic, 2010). Moreover, Callaway *et al.* (2008) stated that the "link between human salmonellosis and host animals is most clear in poultry" and that raw eggs and undercooked poultry are considered by the entire community to be hazardous. Eggs have been implicated as vehicles in numerous outbreaks of salmonellosis; in particular, eggs have been a major vehicle of transmission of strains of *Salmonella* serovar Enteritidis, though the incidence of disease associated with this particular mode of transmission has decreased dramatically (Braden, 2006).

There have been large increases, in Australia and the United States of America (6.5 times), in *per capita* consumption of poultry products since 1910, compared with modest increases (20%) in consumption of beef and pork (Buzby and Farah 2006; ABARE, 2008). Chicken meat is close to replacing beef as Australia's preferred meat, largely because of the industry's success in reducing real (net of inflation) supply costs and hence price to consumers, and through continuous product innovation. The chicken meat industry has reduced supply costs by consistently raising both on-farm and off-farm productivity over several decades through a combination of better management, genetic improvement, economies of scale and mechanisation in processing (ABARE, 2008).

With projected increases in poultry consumption and the association of *Salmonella* with poultry products, the industry needs to employ specific and economic strategies to minimise risk to public health. The production of commercial poultry uses a pyramidal multiplier structure, with breeders at the apex and broilers at the base.

The two broad approaches to *Salmonella* control are 'top down' and 'bottom up', with the latter used widely in Australia. This approach uses spin chiller chlorination within the processing plant for reduction of microbial populations. This method has limited success in reduction of *Salmonella* prevalence and, with consumer choices moving towards 'natural' (chemical free) products, has lead to experimentation with and application of 'top down' intervention strategies. A 'top down' approach, using vaccination at breeder level, can be very cost-effective and is a proven method of flock protection for both viral and bacterial pathogens such as Infectious Bursal Disease Virus, Inclusion Body Hepatitis and *Mycoplasma* (Marangon and Busani, 2007).

The increasing costs or impracticality of improvements in biosecurity, hygiene and management, coupled with the increasing problems associated with antibiotic resistance, suggests that vaccination in poultry will become more attractive as an adjunct to existing control measures (Zhang-Barber *et al.*, 1999).

Although *Salmonella* is typically not a pathogen in the gut of the chicken, systemic infection can cause serious disease (fowl paratyphoid) in the bird (Lutful Kabir, 2010). This usually results from contamination of egg shells from infected breeder hens and spread of the organism into the respiratory tract during hatching which may result in potentially high morbidity and mortality in young chicks (Lutful Kabir, 2010).

Salmonella vaccines available to the poultry industry are based on inactivated or killed cells or genetically attenuated strains (Methner *et al.*, 1999; Babu *et al.*, 2003). However, with concerns in Australia over the use of genetically modified organisms in the food chain, until 2008, only inactivated vaccines were available for use (Anonymous, 1999).

Vaccination against *Salmonella* was first demonstrated to be successful in decreasing poultry mortality due to the poultry-specific *Salmonella* serovar Gallinarum (Smith, 1956). This success, and the Enteritidis disaster in table eggs during the late 1980s that occurred in Europe and the United Kingdom, lead to the development of a *Salmonella* Enteritidis vaccine, which reduced the prevalence of the serovar in poultry. This was followed by a combined (dual) vaccine containing serovars Enteritidis and Typhimurium that again aided in the reduction of carriage in flocks (Clifton-Hadley *et al.*, 2002; Woodward *et al.*, 2002).

Van Den Bosch (2003) reported success with inactivated Salenvac® vaccines in the European poultry industry, which has been associated with a decline in the presence of *S.* Enteritidis and Typhimurium in England. Salenvac®T is a bivalent *S.* Typhimurium and *S.* Enteritidis bacterin-based vaccine, developed from the monovalent Salenvac® *S.* Enteritidis bacterin (Clifton-Hadley *et al.*, 2002).

Based on these studies, production of a trivalent vaccine was initiated in 2004 in association with Intervet Schering-Plough Australia Pty Limited, incorporating the three most predominant *Salmonella* serovars, Typhimurium (group B), Mbandaka (group C) and Orion (group E), from a commercial poultry meat producer, determined through in-house analysis of environmental samples from breeder houses.

Australian poultry companies need to ensure that their operations are free of the specific serogroup D serovars that are present in poultry, including the host-specific chicken pathogens, Gallinarum and Pullorum (Barrow et al., 1994) as well as Enteritidis (Arzey, 2005), a serovar of significant public health concern. This is achieved by ongoing monitoring of hatcheries, feed mills and poultry rearing houses, and culling all positive flocks; therefore these serovars are no longer endemic, with only sporadic cases in Australian broiler and layer flocks (Davos, 2007).

On the other hand, *Salmonella* Typhimurium, which is associated typically with at least 35% of human cases of salmonellosis in Australia, is present in 14% of broiler and 17% of layers isolates (Davos, 2007). Furthermore, other predominant *Salmonella* serovars isolated from broilers and layers include Sofia (38 % broiler, 0% layer), Infantis (10% broiler, 8.9% layer), Montevideo (8.8% broiler, 5.7% layer), Muenster (7.8% broiler, 0% layer) and Mbandaka (<1.7% broilers, 10.5% layers), (Davos, 2007).

The vaccinated or naturally challenged hen protects her progeny through maternal antibodies (IgY) transported into the egg yolk (Kowalczyk et al., 1985). These maternal antibodies offer some protection to the hatchling until its own immune system is fully functioning (Kowalczyk et al., 1985). The IgY suspended in egg yolk can be easily extracted via chemical methods, and purified (Ko and Ahn, 2007). Therefore *Salmonella* vaccinated hens can transfer to the yolk IgY that can be extracted, measured and used prophylactically (in feed) to protect young chicks from initial *Salmonella* colonisation (Vanderplas et al., 2010; Cox and Pavic, 2010).

The objectives of this research were to develop an autogenous tri-vaccine to prevalent poultry serovars, test whether heterologus protection is possible, and if specific anti-*Salmonella* IgY, extracted from the eggs from vaccinated hens, confers prophylactic protection to day old chicks.

2. Material and methods

The methods summarised in this section are divided into three sections, with general, vaccine trial and prophylactic methods individually described.

2.1 General methods

The methods summarised in this section describe the common techniques that were used throughout the experimental work. These methods include: the *Salmonella* strains, challenge suspension and isolation testing; blood and yolk collection and their subsequent testing with the anti-Typhimurium ELISA; trial farm setup and the animal ethics requirements.

2.1.1 *Salmonella* strains

All *Salmonella enterica* subsp. *enterica* strains used in this research (Table 1) were from Birling Avian Laboratories Reference Collection, isolated from the field (production environment), and typed by the Institute of Medical and Veterinary Science (IMVS) in accordance with the White-Kauffmann-Le Minor scheme (Grimont and Weill, 2007).

Serogroup	Antigenic structure	Serovar	Abbreviation
B	1,4,[5],12 f,g,s [1,2] [z27],[z45]	Agona	SA1
B	1, 4 ,[5],12 i 1,2	Typhimurium	ST12
C	6,7,14, r 1, 5 [R1..], [z37], [z45], [z49]	Infantis	SI1
C	6,7,14 z10 e,n,z15 [z37],[z45]	Mbandaka	SM1
E	3, {10}, {15, 34}, y 1, 5	Orion	SO1
E	3,{10}{15} k 1,5	Zanzibar	SZ1

Table 1. *Salmonella* challenge strains belonging to serogroups B, C and E with their respective abbreviation used in the text.

Pure cultures on nutrient agar (NA, Oxoid Thermo Fisher, CM3) were harvested with a cotton swab (Copan, Ref 8155CIS, Italy.) and preserved (10 per serovar) in Cryovials (PRO-LAB Diagnostic, REFPL.170/M, Ontario, Canada) stored at -70°C.

2.1.2 *Salmonella* suspension for challenge

For each *Salmonella* strain, a bead from a Cryovial (PRO-LAB Diagnostic, REFPL.170/M, Ontario, Canada) was incubated in 100 mL of buffered peptone water (BPW, Oxoid Thermo Fisher, CM509, Hampshire, UK) to produce a seed culture. Purity of the culture was checked on NA and identity confirmed serologically using antisera (PRO-LAB Diagnostic Ontario, Canada; Refs TL6002 [O], TKL6001 [H], RL6011-04 [B], PL6013 [C] and PL6017 [E]).

After purity was confirmed and serology determined, isolated colonies were selected and suspended in BPW (4 mL) to give a 75% transmittance (1.0 McFarland) equating to 2×10^8 cfu / mL (bioMérieux, 47100-00 DR 100 Colorimeter, Marcy l'Etoile, France). The target dosage required was achieved through decimal dilution, with the target dilution used as inocula in challenge or recovery experiments, in accordance with AS/NZS5013.11.1 (2004) and confirmed by spread plate enumeration on SM®ID₂ (bioMérieux, Ref 43621, Marcy l'Etoile, France).

2.1.3 *Salmonella* isolation, confirmation and serological testing

All testing was performed at a laboratory accredited by the National Association of Testing Authorities (Australia) in accordance with ISO 6579:2002. The *Salmonella* serovars, listed in Section 2.1.1, were isolated from commercial poultry houses (from visceral emulsions, drag swabs and faeces) on the Eastern seaboard of Australia.

Samples were initially emulsified 1:10 in BPW and incubated at 37 °C for 24 h (standard incubation temperature unless otherwise stated). Aliquots of 1000 µL and 100 µL were transferred into selective Muller Kauffman (MK, bioMérieux, Ref 42114, Marcy l'Etoile, France) and Rappaport-Vassiliadis (RV, bioMérieux, Ref 42110) (incubated at 42 °C for 24 h) broths respectively.

The following validated modifications to ISO 6579:2002 were used. The selective-differential plating agars were Hektoen and XLD (Oxoid Thermo Fisher, PP2027, Adelaide, Australia) and suspect positives (black colonies with clear edges) were confirmed on chromogenic SM®ID₂ instead of the standard biochemical tests (urease, sorbitol fermentation and iron agar reactions).

Presumptive *Salmonella* were serologically confirmed with poly-O and poly-H antisera (Pro-Labs Diagnostic, Refs TL6002 and TKL6001, Ontario, Canada) after subculture onto two slopes of NA and employing the slide agglutination technique. The confirmed *Salmonella* isolates (each on a NA slope) were forwarded to the Australian *Salmonella* Reference Laboratory at the Insitute of Medical and Veterinary Science (IMVS) for complete serological and phage typing.

2.1.4 Blood collection

Blood was collected using a 21-gauge needle (Greiner bio-one, 450072, Germany.) into a 2 mL serum collection tube (Vacuette, 454096, Greiner bio-one, Austria) and, after clotting, transported to the laboratory in chilled containers. The serum was decanted into 1200 μL plasma tubes (Scientific Specialist, 1750-00, CA., USA) and the presence of antibodies to the vaccine was determined using a commercially available *Salmonella* Typhimurium ELISA kit (Guildhay, {trading as x-OVO since 2008, Castle Court, UK}, Flockscreen™, Cat. No V020-43308), according to the manufacturer's instructions (2.1.7).

Initially, the serum was diluted (1:500) by adding 5 μL of serum into 2.5 mL reconstituted sample diluent (2.1.7) in plastic 5 mL tubes (Techno-Plas, 10255001, SA, Aust) and inverted a total of three times to mix.

2.1.5 Egg yolk collection

Eggs were collected after laying by mature hens and transported whole to the lab where they were broken and the yolk separated from the white using a domestic egg separator. The yolk was then poured into a 70 mL sterile jar (Techno-Plas, 10431011, SA, Aust) from which a 200 μL aliquot was removed and added to 1.8 mL of reconstituted wash buffer and mixed by repeated (5 x) aspiration. The yolk was further diluted (1:50) by pipetting 50 μL into 2.5 mL of reconstituted sample diluent buffer (2.1.7).

2.1.6 Immunoglobulin Y extraction

Only Typhimurium ELISA-positive or suspect (2.1.7) egg yolks from Section 2.1.5 were used for IgY extraction. The method selected was the water dilution method described by Staak *et al.* (2001). Initially, the weighed ELISA positive egg yolks were diluted 1:5 w/v with distilled water, mixed vigorously by vortexing (15 s) and frozen at -20 °C for 72 h. Post-freezing the yolk suspension was thawed slowly in a refrigerator at 4 °C.

The thawed yolk water suspension was transferred into centrifuge tubes (50 mL) (Greiner®, T2318, Sigma-Aldrich, St Louis, MO, USA) and centrifuged (Eppendorf, 5810R, Hamburg, Germany) at 2,800 x g (Equation 1) for 20 minutes at room temperature.

$$RCF = 1.118 \times 10^{-5} r N^2$$
$$g = r(2\pi N)^2 / RCF$$

RCF = Relative centrifugal force.
r = rotation radius in centimetres.
N = revolution per minute
g = gravitational force
π = *Pi*

Equation 1. Calculation of centrifugal g forces.

Post-centrifugation, the supernatant was decanted into a volumetric cylinder and the precipitate discarded. To each millilitre of supernatant, 0.27 g of ammonium sulphate ('salt') was added, mixed by vortexing (15 s) and incubated at room temperature for 2 h. Post-incubation the 'salt'-yolk suspension was centrifuged as mentioned previously and the supernatant was discarded.

The precipitate containing IgY was resuspended, in 24 mL of ammonium sulphate (2 M), vortexed (15 s) and incubated at room temperature for 40 min. After incubation, the salt-yolk suspension was re-centrifuged and the supernatant discarded. The final precipitate of crude IgY was resuspended (vortex 15 s) in 5 mL of phosphate buffered saline, transferred to a 10 mL sealable test tube (Techno-Plas, 10281003, SA, Aust.) and stored at 4 °C.

2.1.7 Typhimurium ELISA method

The ELISA was performed in accordance with the manufacturer's instructions (Guildhay, Castle Court, UK). Briefly, kits were allowed to reach room temperature and the wash buffer (100 mL {phosphate buffer with ProClin 0.63 % v/v} to 1900 mL deionised water) and sample diluent (100 mL {phosphate buffer with protein stabiliser and ProClin 0.63% v/v} to 900 mL deionised water) prepared.

Into individually pre-coated (Typhimurium somatic liposaccaride antigen) ELISA strips containing eight wells, 50 µL was dispensed of positive controls (x2), followed by negative controls (x2) and finally the test sample(s). The ELISA plate(s) were covered with an adhesive plastic film (Sealplate, 100-seal-PLT, Excel Scientific Inc, Victorville, CA, USA), mixed by gently tapping the side and then incubated at 37 °C for 30min.

Post-incubation, after removal of the adhesive cover, each well was washed four times (100 µL per well/per wash cycle) in a pre-programmed ELISA Plate washer (Immunowash, 1575, BioRad, CA, USA) with the reconstituted wash buffer. After washing, the plates were dried (five firm taps) by inverting them over paper towel. Once dry, 50 µL of antibody-enzyme conjugate (donkey anti-chicken IgG {Guildhay, Castle Court, UK}) was pipetted into each well. The plate was resealed, mixed and incubated as described previously.

Following incubation, the wells were washed as above, then 50 µL of ELISA substrate reagent (alkaline phenolphthalein monophosphate and enzyme co-factors in a diethanolamine buffer) were added, and the plates covered, mixed and incubated at 37 °C for 15min. The final step was to add 50 µL of ELISA stop solution (1 M sodium hydroxide), ensuring that any bubbles formed were removed, then the plates were analysed in a blanked Microtitre Plate Reader (Vmax Kinetic microplate reader, Molecular Devices, CA, USA) at λ550 nm.

The Guildhay ELISA can detect anti-Typhimurium antibodies, within serum or egg yolk, at an initial dilution of 1:500. After the optical density (OD) all wells was measured at λ550 nm, the sample OD was compared to the mean positive controls OD (Equation 2) to produce the Sample / Positive (S/P) ratio. Using Equation 3 the S/P ratio was converted into a titre.

$$\frac{\text{Sample Optical Density} - \text{mean Negative Optical Density}}{\text{Mean of Positive Optical Density} - \text{mean Negative Optical Density}}$$

Equation 2. Sample to Positive ratio calculation.

$$\log_{10} \text{titre} = 1.046 \times (\log_{10} S/P) + 3.524$$
$$\text{Titre} = \text{antilog of } \log_{10} \text{titre}$$

Equation 3. *Salmonella* Typhimurium titre calculation from S/P ratio.

The Typhimurium titres were initially calculated manually, though software (Guildhay, Castle Court, UK) is available that performs all the calculations and reports the values as OD, SP or titres. According to the kit manufacturer (Guildhay) the lowest threshold for a positive was an SP ratio > 0.25 (titre > 785, OD > 0.173), negative SP < 0.15 (titre < 459, OD < 0.15) and a suspect band SP 0.15 to 0.25 (titre 460 to 784, OD 0.15 to 0.173).

2.1.8 Trial farm set up

The trial house used was an insulated broiler house with side curtains for ventilation control, equipped with two gas-fired space heaters for brooding, and internal fans and roof sprinklers for cooling. The house was divided into 32 individually numbered floor pens, 2.5 x 3 m, with individual bell drinkers and two tube feeders.

To minimise cross-contamination and identify the presence of non-inoculant *Salmonella* the following procedure was used in all trials. Prior to the trial all pens were disinfected using a synthetic phenol (Farm Fluid™, Antec International, Suffolk, UK), drag-swabbed and tested for the presence of *Salmonella*. Fresh litter (wood shavings) was spread evenly across each pen and drag-swabbed again. A footbath, containing a commercially available iodophor-based sanitizer (Sanichick™, Ranvet, Sydney, Australia), was placed outside each of the pens. An empty pen was left between all populated pens, with controls located furthest from the entrance; all routine maintenance started from the controls and worked backwards. Vermin baits, pest strips and an Insectocuter were placed around the house.

The pens were divided into treatment sets, i.e. vaccination, (north side of house) and control, i.e. non-vaccinated (south side of house), with feed and water supplied *ad libitum*. Disposable overalls (Fabri-cell, 05250XL, Vic. Aust.), dust masks (3M, 9320, UK) and gloves (Livingstone Int, GLVLPF100LG-T, NSW, Aust.) were worn inside the house at all times and changed between treatment and control groups. Biohazard bags (Bacto, BCWB66112, Sydney, Aust.) were used to remove contaminated waste, litter and to transport culled or naturally deceased chickens. Hands were washed with an iodophor-based disinfectant, prior to leaving the house. Post-trial the house was disinfected with phenolic-based chemicals and drag-swabbed.

At the termination date, the chickens were euthanized by lethal intraperitoneal injection with 0.5 mL/kg pentobarbitone (Lethabarb®, Virbac Pty Ltd, 1PO643-1, Carros Cedex, France). The carcasses were transported in biohazard bags back to the laboratory and the caeca were removed aseptically by a veterinarian, cut into ten pieces and placed into a single sterile 250 mL sample jar (Techno-Plas, 10453003, SA, Aust.), diluted as per specific procedure with BPW, then incubated for detection as described in Section 2.1.3.

2.1.9 Animal ethics

The Birling Animal Ethics Committee (BAEC), supervised all experimental work in accordance with the Animal Research Act of NSW (1985) and Regulations (2005), following the NHMRC (National Health and Medical Research Council) guidelines (2008) and

NHMRC/ARC (National Health and Medical Research Council, Australian Research Council and Universities Australia) Code of Conduct (2007). When the project was approved, it was designated with a unique BAEC number and a time period for completion.

2.2 Vaccine trial

This section describes the process from the original controlled animal pen to field trials, which occurred over a period of three years.

2.2.1 Vaccine manufacture and administration

Intervet Schering-Plough Australia was commissioned to produce an autologous trivalent inactivated vaccine using proprietary Salenvac® technology from poultry field isolates of Typhimurium, Mbandaka and Orion. Strains were grown on iron-depleted agar, improving specifically the expression of the antigenic iron regulatory proteins (IRPs), further stimulating the humoral response and increasing antibody titres (Van Den Bosch, 2003). The trivalent vaccine was produced using equal amounts of cell suspension (3×10^8 cfu / mL), combined with an aluminium hydroxide adjuvant, and administered to hens intramuscularly into the breast, at 12 and 17 weeks of age.

2.2.2 Experimental animals

In Experiment 1 a total of 50 vaccinated and 50 non-vaccinated, 20-week-old Cobb breeders were placed into trial pens at a minimum of 12 per group, with the remainder used as negative controls. These were obtained from a commercial broiler breeder farm where the vaccine regimes were administered under commercial conditions. All birds where individually labelled with leg or wing tags and blood and rectal swabs collected and tested five days prior to challenge.

In Experiment 2 a total of 100 non-vaccinated, 12 week old Cobb breeders were placed into the trial pens (12 per group) as mentioned previously. In Experiment 3 Cobb broiler day old chicks ($n = 100$) were sourced from a commercial hatchery from vaccinated ($n = 50$) and non-vaccinated parents ($n = 50$).

2.2.3 Adult hen challenge design and trial (experiment 1 and 2)

The number of birds required for each experiment was calculated using statistical tables as described by Martin et al., (1988) to determine the lowest number of repeats required to demonstrate a difference of 20% prevalence between vaccinated (expected 30% prevalence) and non-vaccinated (expected 50% prevalence) groups, at 90% confidence.

As this study involved a new vaccine, the standard deviation of titres following vaccination was unknown. Therefore, 12 birds per group allowed estimation of the average result within a bound of 0.5 x standard deviation from a flock of >300 birds, at 90% confidence (Hancock et al., 1988).

Each trial required the use of only half the trial house (2.1.8) and therefore the other half was sectioned off. The controls were placed furthermost away from the entrance to minimise any cross-contamination. The challenge groups were placed in pens that were opposite each other. Each bird was challenged with 250 µl of the 10^7 cfu *Salmonella* suspension via oral gavage using a 2 mL variable volume pipette (Finnpipette stepper, 4540000, Thermo Electron Corp. Waltham, MA, USA).

The initial experiment (Experiment 1) involved autologous challenge (*i.e.* with the parent vaccine strains) and the subsequent experiment (Experiment 2) involved heterologous challenge, using alternative, poultry-associated serovars from the same respective serogroups as the vaccine strains: Agona (SA1, serogroup B); Infantis (SI1, serogroup C); and; Zanzibar (SZ1, serogroup E). An additional group, challenged with Typhimurium, was used to show repeatability. Cloacal swabs were taken at days 0, 3 and 14 post-challenge. At day 21 post-challenge each bird was bled, humanely euthanized and their caeca removed for culture.

Prior to the heterologous (Experiment 2) trial, the hens were sourced as 10-week old birds, prior to vaccination (12 weeks of age), from a commercial broiler breeder farm that was *Salmonella*-free (confirmed by testing of drag swabs and faeces). These hens were individually tagged and bled, then hand-vaccinated via intra-muscular injection and bled again at 14, 16, 18 and 20 weeks of age.

2.2.4 Progeny challenge trial (experiment 3)

In this trial, 100 chicks were obtained from a commercial hatchery, from eggs produced by vaccinated ($n = 50$) and non-vaccinated birds ($n = 50$). The chicks were vaccinated according to current broiler practices (Marek's Disease, Infectious Bronchitis and Newcastle Disease). Chicks were identified as to their dams' vaccination status by toe web marking and were placed in the trial house at 10 chicks per pen. Pen dividers were used to keep birds close to feed and water, with 2 pens of chicks for each progeny group and challenge (total of 10 pens).

At arrival, blood samples were collected from 12 euthanized (cervical dislocation) chicks of vaccinated and non-vaccinated groups, for *S.* Typhimurium antibody assay (2.1.7). The paper from each box of chicks delivered to the farm was cultured for *Salmonella* as described in Section 2.1.3. The remaining chicks were challenged with *S.* Typhimurium ST12, 10^4 or 10^8 cfu per bird by oral gavage, with controls receiving the diluent buffered peptone water

On days 0, 3 and 14, cloacal swabs were collected, from a random sample of five birds per pen, for individual *Salmonella* culture, with blood samples collected via wing bleeding, on days 7, 10 and 14. On day 21 all birds were bled prior to being euthanized by lethal injection (Lethabarb) and their caeca removed for *Salmonella* testing. All birds were weighed on days 7, 14 and 21.

Any bird that appeared sick, as described in the bird health monitoring sheet, was euthanized immediately, weighed, necropsied and the caeca cultured for *Salmonella* (2.1.3).

2.2.5 Serology

Blood was collected, using a 20-gauge needle, into a serum collection tube and, after clotting, transported to the laboratory in containers chilled with ice bricks. The presence of antibodies to the vaccine was determined using a commercially available *Salmonella* Typhimurium ELISA kit, according to the manufacturer's instructions (2.1.7).

An additional commercial antigen-based ELISA (Idexx, Art.Nr. 99-44100, Liebefeld-Bern, Switzerland), for determination of titres against *Salmonella* serovars prevalent in swine (Typhimurium, Infantis and Enteritidis), was sourced and used according to manufacturer's instructions, with the following modification. All serum dilutions were 1:500, the previously mentioned x-OVO Guildhay conjugate, substrate and stop solution were used, and the ELISA plates read at 550nM.

2.2.6 Longitudinal analysis

The prevalence of serovars over time was determined by performing drag swabs on all broiler breeder flocks from two Australian poultry companies, in three states (New South Wales, Victoria and South Australia), which implemented the *Salmonella* tri-vaccine protocol described in 2.2.2. Swabs were taken at 6, 14, 18, 22, 33, 43 and 53 weeks of age. The *Salmonella* prevalence data were calculated for the years 2003 (pre vaccination), 2004 (during vaccination) and 2005 (post vaccination).

These data (2003 to 2005) were analysed initially for annual prevalence (total positives / total samples received) and the monthly prevalence was calculated by dividing the monthly positive by total tested. This same data set was used to calculate the age based prevalence and determine the serovars present.

The flocks mentioned previously were also bled (2.1.4) regularly (22, 32, 42 and 52 weeks of age) and the serum (*n* = 12) was tested for anti-Typhimurium antibodies using ELISA (2.1.7). From these data the flock immunity, the number of positive sera from the total tested (*n* = 12), and the flock mean was calculated. The flock serum data were sorted by flock age (weeks) and descriptive statistics (mean, medium, standard deviation and 95% confidence limits) were calculated.

The final analysis was to compare the effects of vaccination upon Typhimurium colonised chicks (<12 weeks of age). This analysis was performed by reviewing flock data (Typhimurium antibodies titre and corresponding drag swabs) that contain Typhimurium colonised birds, based on the whole of life cycle throughout 2004/2005. These colonised flocks were sorted upon age (weeks) with the corresponding serology and drag swab data added. A flock was deemed negative for Typhimurium if it had two consecutive negative results (i.e. 18 woa positive, 22 woa negative, 33 woa negative then the flock was deemed negative at 22 woa) and that initial age was plotted against the flock mean Typhimurium titre. These data were also used to evaluate if there was a statistical (χ^2) relationship between high (>1000) and low (<1000) mean antibody titre and flock Typhimurium status at point of lay (25 woa).

2.2.7 Statistical analysis

Qualitative data were converted to numerical data, by assigning (0) for non-detection and (1) for detection, and analysed in a 2 x 2 contingency table, as described by Petrie and Watson (1999). Statistical significance ($P \leq 0.05$) was determined using either the Fisher exact test (any cell with ≤ 5 observations) or the *Chi* squared test (all cells had >5 observations).

The Mantel-Haenszel stratified contingency table test was employed to compare similar treatments stratified across experiments (Thrusfield, 2005). The Student *t*-test was used to evaluate the null hypothesis on vaccination status based on anti-*Salmonella* antibody levels among paired groups of hens. A difference was considered significant at $P \leq 0.05$. All analyses were performed using Statistica™ (StatSoft Inc, 2001, Tulsa, OK, USA).

2.3 Immunoprohylaxis trial

2.3.1 Dried egg yolk preparation

Non-fertile eggs were sourced from a commercial poultry company that routinely administers the trivaccine (2.2.1) to their flocks. The eggs ($n = 400$) from three different farms were sourced from young hens with high (> 1000) anti-Typhimurium serum antibody titres. Upon arrival at the laboratory, 20 eggs were tested for the presence of yolk antibodies to *Salmonella* using the *Salmonella* Typhimurium ELISA kit (2.1.5 and 2.1.7). The remaining egg yolks were pooled into approximately 100 g lots and homogenised by vortexing for 15 seconds. These pooled eggs were then frozen and freeze dried (Avanti JE, Beckman Coulter, Bree, CA. USA).

The dried egg yolk lots were resuspended (weight/volume) in PBS buffer (2.1.5) and tested for the presence of anti-*Salmonella* Typhimurium antibodies using ELISA (2.1.7). The lots were composited into one container and homogenised by vigorous shaking and tested ($n = 20$) for anti-*Salmonella* antibody, as previously described (2.1.7). The dried egg yolk powder was then stored in an air-tight container until required.

2.3.2 Crude extraction of IgY

The freeze-thaw technique (Staak *et al.*, 2001) with ammonium salt precipitation, described fully in Section 2.1.6, was used.

2.3.3 Provision in feed

The feed was divided into three lots: the first lot incorporated 3% w/w IgY egg yolk (dT-IgY) as described by Gurtler *et al.* (2004). Dried egg yolk powder was mixed through a standard commercial broiler breeder starter ration supplied by a local mill. The second lot included dT-IgY re-composited in water (w : v) and, in the final lot, the dT-IgY was re-composited (w : v) in crude IgY extract.

The dosage of dT-IgY was calculated at 3% of total daily feed intake (1.42 g/chick) from the trial mid-point age (11 days of age) according to breeder specifications (Anonymous, 2007). All the feeds were prepared in 250 mL containers and fed to the chicks. The re-composited dT-IgY was initially smeared onto the beaks of individual chicks and the remainder spread in a straight line on chick paper, prior to supplying the standard ration.

The residual from each of the lots (in 250 mL containers) was weighed and an initial dilution (1 : 10) was made with *Salmonella* Typhimurium ELISA buffer. This suspension was diluted (1:2) in plasma tubes and Typhimurium IgY was tested using the ELISA method (2.1.7). The endpoint titre was converted to titre per gram of feed.

2.3.4 Challenge strain

A cryobead of *Salmonella* Typhimurium (ST12), isolated and prepared as in Section 2.1.2, was used as the challenge strain The target density of 10^4 and 10^5 cfu/mL was achieved by decimal dilution of the colorimetrically confirmed 2×10^8 cfu/mL initial suspension (2.1.2).

2.3.5 Animal trials

The trial farm was prepared as stated in Section 2.1.8. Groups of 20 chicks (non-*Salmonella* vaccinated flocks), individually identified by wing tags, were given one of three lots of feed formulation at one day of age and throughout the trial (15 days). On day 3 post-hatch, each chick was challenged (10^4 or 10^5 cfu/mL) by oral gavage (0.250 mL) with *Salmonella* Typhimurium (2.1.2). Faecal samples ($n = 5$) were collected from each pen on days 3 (pre-challenge), 4, 5, 7 and 14 and cultured for the presence of *S.* Typhimurium (2.1.3). The birds were individually weighed at days 0, 7 and 14. At 15 days of age (doa), all birds were humanely euthanized and their caeca removed for culture and enumeration (2.1.3 and 2.3.6).

2.3.6 Enumeration of *Salmonella*

The method employed to enumerate caecal salmonellae was the miniMPN as described in Pavic *et al.,* (2010). Briefly, the removed caeca were cut into sections, to which BPW (w : v) was added (10^0 dilution). A millilitre of this dilution (10^0) was added to a 1250 µL plasma tube and subsequent decimal serial dilutions were prepared (100 µL into 900 µL) in plasma tubes.

Into appropriately labelled microtitre trays, 100 µL of each dilution was added into the corresponding well using a multi-channel pipette. This resulted in the formation of a 3-tube MPN, which was covered with a plastic film (SealPlate®, Excel Scientific, Inc, Victorville, CA, US) and incubated at 37 °C for 24 h. Post-incubation, 100 µL from each microtitre well was added to 200 µL of modified semi-solid RV (MSRV) via a multi-channel pipette and incubated at 42 °C 24 h.

All pale/white wells post incubation was confirmed using SM®ID$_2$ and typical colonies were confirmed serologically with Poly O, poly H and anti-serogroup B antisera. The confirmed data set was converted to cfu/mL, using the MPN charts produced by the United States FDA (2006) from the 3 lowest positive dilutions, and calculated to MPN (cfu) per gram of caeca.

3. Results

3.1 Vaccine trial

3.1.1 Autologous (experiment 1) and heterologous (experiment 2) challenge trials

A challenge with *S.* Typhimurium, used in Experiments 1 and 2 (Table 2) to demonstrate repeatability, showed a significant difference (Mantel-Haenszel Stratified *Chi* squared $P <$ 0.05) between non-vaccinated (colonisation rates of 25% and 50%) and vaccinated (colonisation rates of 0% and 9%) hens. *S.* Typhimurium was also used in both experiments to evaluate seroconversion to the Typhimurium component of the vaccine (Table 4 and 5). The vaccinated flocks exhibited significantly higher (Student *t*-test $P < 0.05$) titres, with 16% and 33% of blood samples having titres 85-6570 (>785 = kit positive threshold), while non-vaccinated hens gave titres of 27-176 (<460 = kit negative threshold).

After challenge, the rates of caecal colonisation (Table 2) in the non-vaccinated hens were 25%, 58% and 17% for serovars Typhimurium, Mbandaka and Orion respectively (Experiment 1), with an average colonisation rate of 33%. In the heterologous trial

(Experiment 2) non-vaccinated hens had an average caecal colonisation rate of 42% calculated from the individual caecal colonisation rates: Typhimurium (50%); Agona (92%); Infantis (17%) and Zanzibar (9%).

Experiment number	Vaccine status	Detected	Not detected	Relative risk	Positive (%)
(a) Typhimurium					
1	Vaccinated	0	12		0
	Non-vaccinated	3	9	undefined	25
2	Vaccinated	1	10		9
	Non-vaccinated	6	6	0.18	50
M-H analysis	Crude OR = 0.08; 95% CI = 0 to 0.69; OR = 0.06; X^2 = 5.73; P = 0.017				
(b) Autologous Typhimurium and heterologous Agona [B]					
2	Vaccinated	1	10		9
	Non-vaccinated	6	6	0.18	50
2	Vaccinated	5	7		42
	Non-vaccinated	11	1	0.23	92
M-H analysis	Crude OR = 0.15; 95% CI = 0.1 to 0.49; OR = 0.08; X^2 = 8.83; P = 0.003				
(c) Autologous Mbandaka [C] and heterologous Infantis [C]					
1	Vaccinated	1	11		8
	Non-vaccinated	7	5	0.02	58
2	Vaccinated	0	10		0
	Non-vaccinated	2	10	Undefined	17
M-H analysis	Crude OR = 0.08; 95% CI = 0.01 to 0.62; OR = 0.05; X^2 = 6.24; P = 0.013				
(d) Autologous Orion [E] and heterologous Zanzibar [E]					
1	Vaccinated	0	12		0
	Non-vaccinated	2	10	Undefined	25
2	Vaccinated	0	10		9
	Non-vaccinated	1	10	Undefined	50
M-H analysis	Crude OR = 0; 95% CI = -; OR = undefined; X^2 = 1.31; P = 0.25, All remaining hens were used as negative controls.				

Table 2. Mantel-Haenszel stratified contingency table analysis of *Salmonella* caecal culture results comparing non-vaccinated and Tri-valent vaccinated Cobb™ adult breeder hens challenged (10^7 cfu / mL) with serovar (CI = confidence Interval; M-H = Mantel-Haenszel; OR = odds ratio).

The cloacal swab results (Table 3) suggested that no hens were colonised by *Salmonella* prior to challenge (Day 0). Three days post challenge (Day 3) the prevalence in non-vaccinated hens ranged from zero to 40 % compared to zero to 10 % for the vaccinated flocks, changing to 0-83 % and 0-36 % respectively 14 days post-challenge. The cloacal swabs (Table 3) showed that the serovar Agona colonised non-vaccinated hens the best, with prevalence of 40 % (at day 3) and 83 % (at day 14), and a caecal prevalence of 92 %. In contrast, Zanzibar did not appear to colonise (day 3 and 14), based on cloacal swabs, and a caecal prevalence of only 9%.

Due to low colonisation rates (17 % and 9 %), combined with the low number of replicates (n = 12 and 11), statistically valid ($P \le 0.05$) results could not be obtained for serogroup E (Table 2d). The null hypothesis could not be rejected. The statistically desired colonisation rate for non-vaccinated (50 %) was only achieved with serovars Mbandaka (Table 2c), Typhimurium and Agona (Table 2b) which all demonstrated a significant difference between vaccinated and non-vaccinated hens.

By using a stratified Mantal-Haenszal *Chi* squared test, comparing autologous to heterologous serovars within the same serogroup, significant differences ($P = 0.003$ and 0.01) were demonstrated for serogroups B (Table 2b) and C (Table 2c), and respective odds ratios of 6.6 and 12.5 times for colonisation among non-vaccinated compared with vaccinated groups. However, serogroup E serovars did not demonstrate a significant difference due to the low colonisation rate of the non-vaccinated hens.

The serovar Typhimurium was used to demonstrate (Table 2a) a repeatable significant difference between non-vaccinated and vaccinated hens (Mantel-Haenszal $P = 0.017$). All non-challenged controls were negative for the challenge and wild serovars of *Salmonella*.

		Prevalence of vaccinated (n =12) / non-vaccinated (n = 12) hens at various days post challenge.			
Exp	serotype	Day 0	Day 3	Day 14	Day 21 caeca
1	Typhimurium	0% / 0%	0% / 16%	0% / 16%	0 % / 25%
1	Mbandaka	0% / 0%	0% / 16%	0% /16 %	8 % / 58 %
1	Orion	0% / 0%	0% / 16%	0% / 16%	0% / 16%
2	Typhimurium	0% / 0%	8% / 25%	9% (11) / 25%	9% (11) / 50%
2	Agona	0% / 0%	10% /40%	36% / 83%	42% / 92%
2	Infantis	0% / 0%	0% / 20%	0% (10) / 8%	0% (10) / 17%
2	Zanzibar	0% (10) / 0%(11)	0% (10) / 0% (11)	0% (10) / 0% (11)	0% (10) / 9% (11)

Prevalence without bracketed numerals were calculated from n = 12 as compared to n = (x). All groups initially contained 12 birds and bird death was due to non-trial related illness.

Table 3. *Salmonella* serovars prevalence from cloacal swabs (0, 3 and 14 days) and caeca (21 days) post challenge (10^7 cfu) from adult Cobb™ breeder broiler hens in Experiments 1 and 2.

3.1.2 Experiment 1 autologous serology

The blood samples taken from all hens prior to challenge demonstrated that vaccinated hens contained variable levels of antibodies to *Salmonella*. This is illustrated by the wide range of titres (1161 ± 623 at the 95 % confidence intervals from mean) from the x-OVO Guildhay *Salmonella* Typhimurium ELISA test (Table 4). All the unvaccinated flocks had titres well below the negative cutoff value. Blood samples from tri-vaccine vaccinated hens (n = 30) were tested, using the x-OVO Guildhay Typhimurium and Idexx Typhimurium-Infantis-Enteritidis ELISA, showing a 1 : 0.8 ratio respectively (x-OVO Guildhay Typhimurium ELISA average OD = 0.39 and Idexx Typhimurium: Infantis: Enteritidis ELISA average OD = 0.70 in positive wells), which was highly significant (Student *t*-test P = 0.0001). All blood sera positive for anti-Infantis antibodies contained anti-Typhimurium antibodies.

Blood serum result	Typhimurium titre mean (95% confidence limits)	Coefficient of variation
Non Vaccinated titre (n = 50)	130 (± 27)	60 %
Vaccinated titre (n = 50)	1161 (± 623)[A]	127 %

[A] Means without common superscripts differ significantly (Student *t*-test <0.05).

Table 4. Anti-*Salmonella* Typhimurium antibody titres means, with 95% confidence limits, for vaccinated and non-vaccinated Cobb™ adult breeder hens (Experiment 1).

The large co-efficient of variation reflects variation in seroconversion among the vaccinated birds, with 16% of those vaccinated demonstrating titres > 785 and 25% with titres > 460 (true negative). There was a statistically significant difference (Student's two-tailed *t*-test, P < 0.05) in anti-*Salmonella* antibody levels between vaccinated and non-vaccinated flocks.

3.1.3 Experiment 2 heterologous serology

Half of the vaccinated group in Experiment 2 were individually tagged and bled prior to vaccination at 12 and 17 weeks of age (Table 5). Hens at 18 weeks had developed significant (Fisher exact P = 0.001) serum antibodies to S. Typhimurium compared to hens prior to vaccination (12 weeks), but not to hens in other age groups. Hens at 20 weeks of age had developed a significant difference in antibody to all age groups (P = 0.0005 and 0.01) with the exception of 18 week olds.

Hen Age (weeks)	Titre <785	Titre >785	Total	Positive (%)
12	25	0	25	0 % [A]
14	22	2	24	8.33 % [AB]
16	22	2	24	8.33 % [AB]
18	17	6	23	26 % [BC]
20	14	9	23	43 % [C]

[AB] Percentages without common superscripts differ significantly (Fisher Exact P < 0.05). Mortalities of hens at 14 and 18 weeks were due to non vaccine-related causes (Femoral Head Necrosis).

Table 5. Anti-*Salmonella* Typhimurium antibody titres in vaccinated hens (n = 25) prior to (12 weeks), during (12 to 17 weeks) and after (18 to 20 weeks) the vaccination protocol (Experiment 2).

After the first vaccination, initial sero-conversion was observed in two (8.33 %) of the hens (n = 24). Unfortunately, one of those hens died at 18 weeks, whereas the other hen had titres of 812, 1222, 1200 and 1114 at weeks 14, 16, 18 and 20 respectively, and this hen was also negative for presence of *Salmonella* in its caeca post-challenge. The percentage 'true' (titre > 785) positive was 26 % or 34 % if the ELISA (2.1.7) kits 'suspect' interpretations were included. When these hens were randomised and challenged at 20 weeks of age the percentage of serum positive birds was 43 %, or 60 %, including 'suspect' reactions. The titre range in the second experiment was more precise with a mean titre of 1016 ± 546(95 % CI) and a low co-efficient of variation (26 %).

The percentage of eggs that had antibodies against *S.* Typhimurium was 16 % (titre > 785), or 48 % (n = 61) if 'suspects' (titre 460 to 784) were included. Titres ranged from 145 to 1890 for egg yolks from vaccinated hens compared with a range of 332 to 427 (below ELISA kit negative < 460) for non-vaccinated hens eggs. Regardless of the low number of ELISA positives (titres > 785) from eggs originating from vaccinated hens, there was still a significant difference (P = 0.001) between egg yolk anti-Typhimurium titres from non-vaccinated as compared to vaccinated hens.

3.1.4 Maternal antibody protection (experiment 3)

All chicks (except non-vaccinated and non challenged control and vaccinated non-challenged control) were challenged at day 0 and appeared healthy and gained weight.

Initial serological testing from a culled group (n = 24) resulted in negative titres (< 460) and chick paper culture was negative for *Salmonella* for both non-vaccinated and vaccinated flocks. There was one chick from the vaccinated 10^8 and non-vaccinated 10^4 (Table 6) which died from conditions unrelated to the trial (Femoral Head Necrosis).

| Groups | 21 day old caecal *Salmonella* | |
Vaccine status / challenge level	Detected	Not detected
[A]Vaccinated / 10^8 cfu	19	0
[A]Non-vaccinated / 10^8 cfu	20	0
[B]Vaccinated / 10^4 cfu	15	5
[A]Non-vaccinated / 10^4 cfu	19	0

The vaccinated / non challenged (n = 10) and non-vaccinated / non challenged (n = 10) controls were all negative for *Salmonella*. Groups without common superscripts differ significantly (Fisher Exact P ≤ 0.05). One chick died from a non-trial related cause (Femoral Head Necrosis) in both the vaccinated 10^8 and non-vaccinated 10^4 group. There were no differences between cloacal swab results from days 3, 14 and 21.

Table 6. *Salmonella* vaccinated (n = 50) and non-vaccinated (n = 50) progeny protection from *S.* Typhimurium challenge at high (10^8 cfu) and low (10^4 cfu) dose.

The cloacal swab results from randomly chosen chicks (n = 5) were all positive 3 days post challenge for *S.* Typhimurium, during the grow-out phase (day 14 swabs) and caecal culture results (day 24) when the birds were challenged with 10^8 cfu *Salmonella*.

The cloacal swabs and caecal cultures revealed a significant difference (Fisher exact P = 0.047) between chicks challenged with 10^4 cfu of *S.* Typhimurium, which were progeny, of vaccinated hens (75% positive) compared to progeny from non-vaccinated (100 % positive) hens at 3, 14 and 24 days of age (Table 6). The serology results, initially and at 21 days post challenge, indicated negative *S.* Typhimurium antibody titres (< 460) for both vaccinated and non-vaccinated chicks.

3.1.5 Longitudinal analysis

The comparison of the prevalence of *Salmonella*, at genus level and within flocks, was performed by plotting the monthly drag swab relative prevalence (positive flocks / total flocks tested in each month) one year prior to vaccination (2003), during vaccination (2004) and subsequent vaccination (2005).

The prevalence of *Salmonella* in 2003 was 52 %, decreasing to 41 % in 2004 and 40 % in 2005 (Figures 1 and 2). The first flocks vaccinated were 12 weeks of age as of the 1/1/2004, designated with an arrow (Figure 1). For each month post vaccination the 'forecast' relative prevalence (Figure 1) was calculated (Excel, forecast function using the previous 12 months prevalence data to calculate the predicted value).

The year prior to vaccination (Figure 1) showed peaks and troughs in flock prevalence, which followed seasons (lowest prevalence in winter and higher prevalence in summer). As the flocks were vaccinated, the monthly prevalence levels decreased and the monthly variation flattened; a moderate linear association (R^2 = 0.56) was found, compared to 'forecast' values (R^2 = 0.16) (Figure 1).

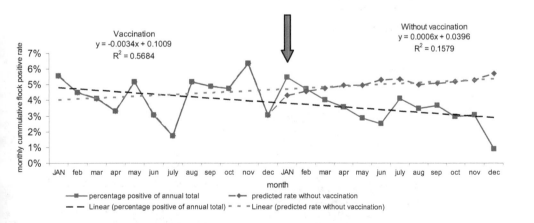

Fig. 1. *Salmonella* flock monthly prevalence (%) in Breeders during the years 2003 (*n* = 794) to 2004 (*n* = 1024) with the arrow indicating when vaccination started and forecast data (Red line) used to show the predicted prevalence if vaccination was not introduced.

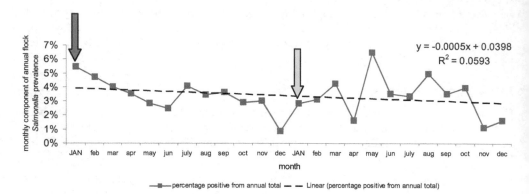

Fig. 2. *Salmonella* flock monthly prevalence (%) in Breeders during the vaccinations years 2004 (*n* = 1024) to 2005 (*n* = 968). The two arrows indicate the start of vaccination and the month when all flocks were vaccinated.

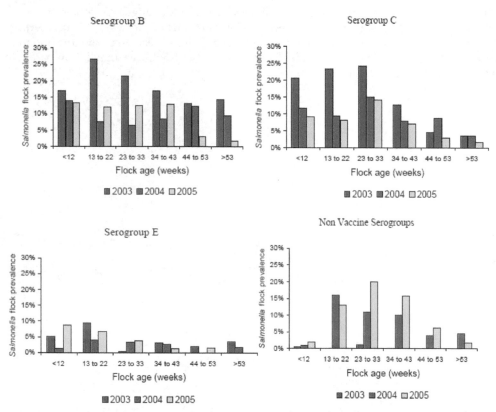

Fig. 3. The prevalence rate in tri-valent vaccine-relevant serogroups in breeders, based upon flock age for the pre-vaccinated year 2003 (*n* = 794) and the vaccinated years 2004 (*n* = 1024) and 2005 (*n* = 968) with vaccination occurring at 12 and 18 weeks of age.

The variation in sample size, 794 (2003), 1024 (2004) and 968 (2005), was due to more intensive testing in the first year of vaccination. The flock prevalence rate between 2004 (40 %) and 2005 (41 %) remained static, which was corroborated by a R^2 (linear) of almost zero (Figure 2). The spikes observed in the latter half of 2005 (Figure 2) were due to the unplanned introduction of non-vaccinated *Salmonella*-positive males, at 20 weeks of age, to other breeder farms.

The results in Figure 3 demonstrate an interesting trend in the prevalence for the years in which the vaccine e was administered (2004 and 2005, blue and green bars) compared to the year prior (2003, red bar). The prevalence of *Salmonella* from serogroups B, C and E decreased post-vaccination while increases among non-vaccine serogroups were observed, with serovar subsp 1 rough:r:1,5 predominating (Table 7).

| | | | Year | |
sero-group	Serovar	2003	2004	2005
B	Agona	15.0%	6.0%	5.8%
B	Subsp 1 ser 4,12:d:-	5.5%	8.0%	0.4%
B	Typhimurium	15.2%	7.2%	9.03%
C1	Infantis	29.2%	26.0%	18.6%
C1	Livingstone	6.1%	0.5%	0.21%
C1	Mbandaka	4.9%	6.0%	6.6%
C1	Singapore	0.6%	0.9%	0.24%
C2	Bovismorbificans	12.6%	3.3%	3.61%
C2	Muenchen	0.6%	0.23%	1.51%
E1	Orion	2.2%	0.23%	0.4%
E1	Zanzibar	3.8%	5.2%	6.2%
E4	Seftenberg	0.6	2.0%	4.7%
	Subsp 1 ser rough:r:1,5	0.6%	23%	14.6%
	percentage total	97.0%	88.56%	71.9%
	total positive flocks	412	420	387

Table 7. Flock *Salmonella* serovar trend for the year prior, 2003 (n = 794), and the years post-vaccination, 2004 (n = 1024) and 2005 (n = 968).

During and subsequent to vaccination there was a change in the profile of serovars isolated from breeder flocks (Table 7). The vaccine serogroups all decreased with the exception of Mbandaka (serogroup C). This increase could be explained by early colonisation prior to vaccination (Figure 4). Table 7 shows serovar succession, which may have occurred post vaccination.

Prior to vaccination (2003) there were 13 serovars that accounted for 97 % of all positives (Table 7). This figure decreased to 88.9 % (2004) and 71.9 % (2005) when all flocks were vaccinated. There was also a major change in the predominant serovar with rough: r: 1:5 increasing from 0.6 % (2003) to 23 % (2004) and 14.6 % (2005), both significantly different (χ^2 P = 0.0001). Beside those in Table 7, the serovars isolated were: in 2003 Lille (1.2 %), Ohio

(0.6 %), subsp. 1 ser 16:I;v;- (0.6 %) and Tennessee (0.6 %); 2004 and 2005 Anatum (0.5 % {2004}, 1 % {2005}), Chester (0.6 %, 1.2 %), Give (1.6 %, 2.0 %), Havana (0.6 %, 5 %), Javiana (0.6 %, 1.5 %), Kiambu (1.5 %, 3 %), Lillee (1.5 %, 1 %), Ohio (1.2 %, 1.9 %), Subsps 1 ser 16:I;v;- (1.5 %, 3.5 %), Subsps 1 ser 6,7:r:- (0.6 %, 0 %), Worthington (1.6 %, 0 %), Bredeney (0 %, 2 %), Cubana (0 %, 3.5 %) and Tennessee (0 %, 2 %).

There was no overall change in prevalence of serogroup C1 (Table 7); while the C1 positive flocks were all positive prior to vaccination at a steady 5 % prevalence rate, there was a shift in serovar prevalence, with a decrease in Infantis, the vaccine-homologous serovar.

Results suggest marginal heterologous serovar protection as shown by the steady decrease in Agona (serogroup B) post-vaccination. This decrease was similar to that for Typhimurium, observed over the same time period, and may explain the reduction in total prevalence observed during the field trial (Figure 1).

The seroconversion of birds to the vaccine Typhimurium component was routinely measured (ELISA) at the ages mentioned in Figure 4, and the mean (n = 12), with 95 % confidence limits, was recorded. The green (line) and the red (line) indicated the Typhimurium ELISA positive and negative titres levels respectively (Figure 4).

The ELISA results (Figure 4) showed a high degree of variability in seroconversion, indicated by the very large error bars (95 % confidence). The level of flock immunity (number of sera tested that had Typhimurium ELISA titres > 785) for each age group (Figure 4) during 2004 (2005) was 23 % (27 %); 15 % (17 %); 11 % (11 %) and 3 % (6.4 %). The average flock immunity for 2004 and 2005 was (titre > 785 positive level) 28 % and 27 % respectively or 40 % and 37 % if the titre was calculated at > 460 (negative level).

The 2005 trend line had a dip at 23 to 32 weeks of age; only 21 flocks (272 sera) were tested in 2004 compared to 135 flocks (1345 serums) tested in 2005, which affected the average. This large discrepancy in flocks tested in this age group (2004 and 2005) was not observed in any other age group during the same period (2004 to 2005).

The titre profile shown in Figure 4 for all flocks was also apparent when the titres of the individual flocks were examined over the same time period (categorized by weeks of age{woa}). They all showed a similar age titre profile (Figure 4; year 2004) with a shoulder period (ages 13 to 22 and 23 to 32) and then a decrease with age.

An investigation into the efficacy of the tri-vaccine on juvenile (< 6 woa) breeder flocks (grown day old to death, rearing and production), already colonized by serovar Typhimurium (n = 32), was undertaken during 2004 to 2005. All these Typhimurium positive flocks (determined using drag swabs) were routinely sampled, as mentioned previously, and the drag swabs were compared to the anti-Typhimurium titre for the same age groups. A flock was deemed negative if two consecutive drag swabs did not contain the Typhimurium serovar, and the initial age recorded (i.e. 6 woa = Typhimurium positive; 12 woa = positive; 18 woa = positive; 22 woa = positive; 33 woa = negative; 43 woa = negative and 53 woa = negative, the age negative is 33 woa).

The corresponding Typhimurium titre was also measured and the flock average over the sampling period was recorded. The data summarized in Figure 5 show a random and variable Typhimurium clearance rate in flocks with low (< 1000) anti-Typhimurium antibody means, with only two flocks clearing the serovar prior to production age (> 25

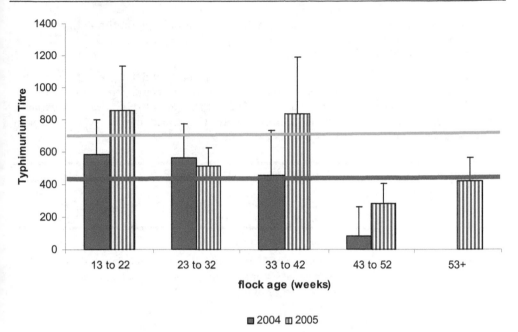

Fig. 4. Flock seroconversion of the Typhimurium component (positive {green line} and negative {red line} ELISA thresholds) of the Tri-vaccine as measure by anti-Typhimurium ELISA (95 % confidence limits) from flock (n = 12) bloods taken at various ages between 2004 and 2005.

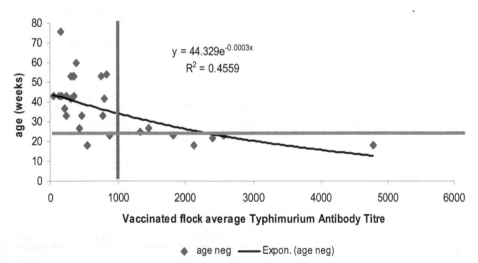

Fig. 5. Naturally Typhimurium colonised chicks (2004 to 2005), grown in day old to death houses, clearance age (weeks) in relation to the flock mean (n = 12) anti-Typhimurium titre trend (exponential R^2), measured post vaccination (12 and 18 weeks), with critical limits set at hen age of 25 weeks (green line) and mean titre > 1000 (blue line).

woa). Six of seven flocks with high titre (> 1000) were clear of Typhimurium at production age (< 25 woa). These flocks with high titre means (> 1000) also had high (75 %, n =12) flock immunity (anti-Typhimurium titre > 850).

The data in Figure 5 were analyzed using an exponential trend line, which showed a moderate association (R^2 = 0.45) that, when converted to a 2 x 2 contingency table (Table 8), was highly significant (χ^2 P = 0.0002) between titre and Typhimurium clearance. Additionally, Typhimurium colonized flocks were 17.5 times more likely to remain colonized after 25 woa, if their titres were below 1000.

Typhimurium titre	Negative at < 25	Positive at >26	Total
High >1000	6	1	7
Low <1000	2	23	25
total	8	24	32

Fisher exact P = 0.0002 with an odds ratio of 140 and a relative risk of 17.5.

Table 8. The *Chi* squared analysis comparing high (> 1000) and low (< 1000) mean (n = 12) anti-Typhimurium titres in pre-vaccinated Typhimurium positive flocks that had consecutive negative drag swabs at 25 weeks of age.

3.2 Immunoprophylaxis trial

3.2.1 IgY analysis

ELISA analysis of the breeder egg yolks from three different farms showed that all the flocks had high flock immunity (number of samples titre > 785 in serum sample divided by total tested) and a high mean flock titre (Table 9). The crude IgY yield (Figure 6) ELISA results (Figure 7) showed a high variance in Typhimurium antibodies (15 % to 80 % > 785 and means of 343 to 2518) deposited in egg yolks tested (n = 20) which was further investigated using Farm 1 yolks (Figure 8).

	Farm 1	Farm 2	Farm 3
Flock Typhimurium serum (IgY) mean (95% CI)	1118 ($\pm$502)	1844 ($\pm$528)	2104 ($\pm$564)
Flock Immunity rate	70 %	85 %	100 %
Egg Typhimurium IgY positive rate	15 %	26 %	80 %
Egg Typhimurium IgY titre mean (95% CI)	343 ($\pm$143)	761 ($\pm$276)	2518 ($\pm$1361)
Crude IgY yield mg/g yolk (95% CI)	28 ($\pm$8)	32 ($\pm$12)	26 ($\pm$4)
Crude Typhimurium IgY mean (95% CI) titre	153 ($\pm$118)	386 ($\pm$286)	690 ($\pm$390)

Table 9. Summary of *Salmonella* Typhimurium vaccinated donor flocks for mean (95% CI) antibody titres, from serum (n = 12), egg yolk (n = 20) and crude extract (n = 20), flock immunity (percentage of sera with titre > 785), yolk positive rate (No. of yolks > 785 titre) and crude IgY yield (mg/g).

Fig. 6. Salt precipitated crude IgY containing Typhimurium antibodies, which were extracted from vaccinated breeder hen's eggs.

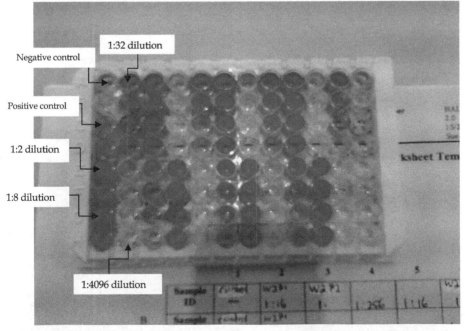

Fig. 7. *Salmonella* Typhimurium ELISA plate, using doubling dilution, to determine the crude IgY titre from egg yolk extract.

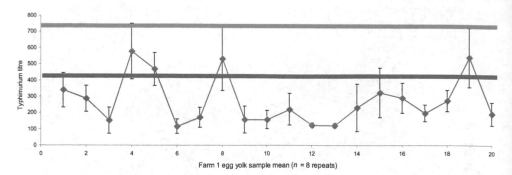

Fig. 8. Variability in anti-Typhimurium ELISA (positive > 785, green line, and negative < 459, red line) titre means (n = 8, 95 % CI) in eggs (n = 20) from Farm 1. The data portrayed in Figure 8 show a very large variance in the Typhimurium antibody levels present in Farm 1 egg yolks. Only egg yolks 4, 8 and 19 had Typhimurium titres levels close to the ELISA positive range of > 785. The majority of yolks, with the exception of yolks 1, 5 and 15, were below the negative level of < 459.

Of the yolks from the three farms, composited post-freeze drying, anti-Typhimurium antibodies (titre > 785) were present in 90 % of the samples tested. The titres in the composites ranged from 16 to 1024 with a mean ($\pm$ 95 % CI) of 386 ($\pm$ 203). The egg yolk composited crude extract (Figure 7) resuspended in PBS (5 mL per yolk), had an IgY titre of 256.

3.2.2 IgY feed trial

The results of the net body weight gain for these four feeding regimes are summarised in Figure 9 which showed that there was no significant difference (Student t-test > 0.05) for all the feeds regardless of Typhimurium challenge dose and at various ages (0, 7, and 14 doa).

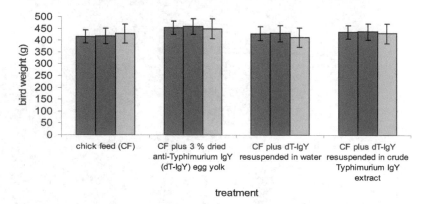

Fig. 9. The net body weight gain (after 15 days) of challenged (10^4 or 10^5 cfu/mL of *S. typhimurium*) chicks fed with four rations.

The rehydrated dT-IgY was fed to the chicks initially by hand (smearing of dT-IgY paste on the beak) with the remainder of the feed spread out in a line on chick paper. Only when all this feed was consumed was the standard ration introduced. This ensured that all the rehydrated dT-IgY was consumed.

The residual treated feed was tested for the presence of Typhimurium antibody. The results, summarised in Figure 10, show that the chicks in the treatment groups received anti-Typhimurium IgY, in various degrees, throughout the trial.

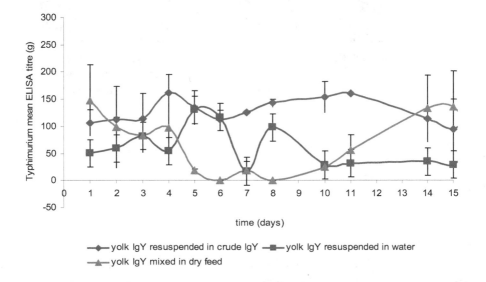

Fig. 10. Typhimurium antibody titre (95% CI) of three treated feed rations: (CF) plus 3% (w:w) freeze dried Typhimurium IgY (dt-IgY) yolk (green line); CF plus dT-IgY resuspended in water (w:v) (red line) and dT-IgY resuspended in crude Typhimurium IgY extract (w:v) (blue line) feed prior (3 days) and post (4 days of age+) Typhimurium (10^4 or 10^5 cfu/mL) challenge.

The data summarised in Figure 10 showed that the dT-IgY resuspended in crude IgY extract (blue line) had the highest and most consistent titres, with a range 80 to 160 and a mean of 110, throughout the trial. The dT-IgY resuspended in water (red line) had the highest titre range of 50 to 130 and a mean of 60. Finally, the dT-IgY egg yolk powder (green line) treatment was the most inconsistent with a titre range of 0 to 150 and a low mean of 61.

The final analysis was to measure the effect of these feed treatments on S. Typhimurium levels in the caeca. The data summarised in Figure 11 show that, regardless of treatment, there was no significant (Student t test > 0.05) reduction in caecal Typhimurium populations as compared to the control. This was also observed in faecal (days 4, 5, 7 and 14) and caecal samples that were 100 % positive when tested qualitatively for *Salmonella* presence.

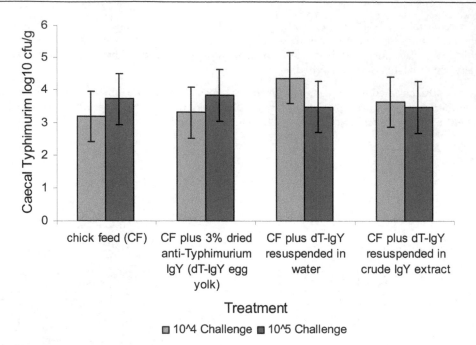

Fig. 11. *S.* Typhimurium caecal counts ($\log_{10}$ cfu/g, 95 % CI) from challenged (10^4 or 10^5 cfu/mL) chicks on three feed treatments.

4. Discussion

4.1 Vaccine trial

One of the objectives behind developing and using a multivalent vaccine was to reduce carriage and shedding of otherwise prevalent salmonellae, and thus to protect the progeny from egg contamination and from colonisation by low level (< 100 cfu) environmental and / or feed exposure (Anonymous, 2004; Franco, 2005; Volkova *et al.*, 2009). The protection of poultry against *Salmonella* infection is largely empirical, based upon experimental trial work and field studies (Anonymous, 2004). Knowledge relating to the course of the innate and adaptive response to colonisation with various *Salmonella* types is beginning to increase (Anonymous 2004; Beal and Smith 2007). To be suitable for use in industry, a vaccine must be safe for poultry and other species, and not interfere with detection of *Salmonella*, as well as effective in eliminating shedding and enhancing clearance (Anonymous 2004).

With the above definition in mind the decision to use the predominant serovar from each of serogroups B, C and E was based upon success of this approach with serovars Typhimurium and Enteritidis in a European study (Van Den Bosch 2003), and serogroups C and E were added primarily to offer protection against the prominent Australian poultry serogroups (Davos, 2005) and, secondarily, to expand the knowledge base (Anonymous, 2004,). There were no serovars from serogroup D included in this vaccine as Australian commercial flocks are free of this serogroup (due to zero tolerance), including serovars Enteritidis, Pullorum and Gallinarum (Arzey, 2005).

According to Grimont and Weill (2007) only 30 serovars account for 90 % of *Salmonella* isolates in any given country. In the Australian broiler industry, typically ten different serovars account for 84 % of all isolates (Davos, 2007). To suit inevitable changes in serovar prevalence, killed vaccines such as the trivaccine in the present study are most suitable for industrial use as they can be more easily prepared and therefore more readily adapted than attenuated strain vaccines.

Prior to 2008, due to consumer fears about genetically modified organisms (GMOs), in the food supply, only killed *Salmonella* vaccines were available for commercial use in Australia (Anonymous, 1999). The next stage of development would include vaccination with killed and attenuated *Salmonella*e to target both cell mediated and humoral immunity.

Bailey *et al.* (2007a) showed increased IgG and IgA production in serum, the gut and crop of adult broiler breeders after administered of a live followed by a killed vaccine, when compared to a killed or attenuated vaccine alone. Attenuated vaccines can be given to day-old chicks, providing competitive exclusion and stimulating polymorphonucleated cell migration to the intestinal cell walls, thus preventing initial colonisation by wild strains (Van Immerseel *et al.*, 2005).

In the present study, a killed multi-serotype vaccine was injected into hens after 12 weeks of age to provide humoral protection. Bailey *et al.* (2007a) administered an *aroA* attenuated Enteritidis vaccine at 1 and 21 days followed by an autogenous killed trivalent vaccine to 11 and 17 week old broiler breeder parents. Following post-vaccination autologous challenge, the reduction in caecal counts and increase in antibody titre were similar in the study of Bailey *et al.* (2007ab) and the present study, suggesting the killed vaccine alone is effective.

However, Deguchi *et al.* (2009) administered an autologous trivalent (Enteritidis, Typhimurium and Infantis) killed vaccine at only 6 weeks of age and was able to show an increased antibody titre and reduced caecal carriage, for challenge (10^9 cfu) with all autologous serovars. The variable results across these studies suggest that the effectiveness of vaccination may vary with the nature and timing of administration of the vaccine. The efficacy of a killed vaccine can be enhanced by using improved adjuvants that may target different parts of the immune system (Barrow 2007).

The modest total colonisation rates in non-vaccinated hens of 33 % and 42 % (Experiment 1 and 2) illustrated that the challenge dose (10^7 cfu) may have been too low or the strains were attenuated. Other studies (Byrd *et al.*, 2003; Babu *et al.*, 2004) showed high rates of colonisation (> 80%) with high challenge doses of Typhimurium (10^8 to 10^9 cfu and 10^{10} cfu respectively). However, the 10^7 cfu challenge dose was chosen as this is a more realistic, field-appropriate, natural challenge, than 10^{10} cfu.

Numerous studies indicate avian susceptibility to *Salmonella* is not uniform and many influencing factors exist. The literature suggests that a stable anaerobic microflora, low house density, bacterial interference, breed resistance, antagonism, colonisation resistance, barrier effects, challenge strain virulence, competitive exclusion and passive immunity could all contribute to low challenge rates (Lloyd *et al.*, 1977; Fuller, 1989; Mead and Barrow, 1990; Kottom *et al.*, 1995; Duchet-Suchaux *et al.*, 1997; Nisbet *et al.*, 2000; Quinn *et al.*, 2000; Chambers and Lu, 2002; Kinde *et al.*, 2005).

The hens used in this study were derived from a field trial situation, stocked at low densities and were healthy, thus decreasing their susceptibility to colonisation to levels less than expected than under the stressful conditions of industry practices. When this vaccine was used under field conditions, over a four year period, the incidence of Typhimurium was reduced in vaccinated flocks with an anti-Typhimurium ELISA titre of > 1000 (Figure 5).

The results summarised in Tables 2.1b and 2.1c highlight autologous and heterologous serotype protection conferred by the multivalent vaccine and these trial results were similar to those of Clifton-Hadley et al. (2002) during trial of the first Salenvac® bi-vaccine. The repeatability with S. Typhimurium in both trials was expected and heterologous protection was demonstrated with the serogroup B serotype Agona. Protection against heterologous serotypes from serogroup B was also demonstrated by Beal et al. (2006).

The serovars Orion and Zanzibar (serogroup E) were both chosen due to their field prevalence at the time in chicken litter and meat (Davos, 2005). However, these serovars both failed to colonise caeca 21 days after challenge. While detection in flocks may suggest colonisation, such serovars may derive from the environment or feed and be transitory or only capable of minimal colonisation in the birds; still detectable by drag swabs of flocks.

The x-OVO Guildhay *Salmonella* Typhimurium lipopolysaccharide ELISA was used in this study to measure the immune response to the vaccine, as it was used during development of the first Salenvac killed vaccine (McMullin et al., 1997). The trivaccine contains 1×10^8 cfu of each of the three serovars added in a ratio of 1:1:1.

The assumption is that the hen develops antibodies to each serovar at titres similar to those determined for S. Typhimurium. Okamura et al. (2007) demonstrated similar *Salmonella* lipopolysaccharide ELISA antibody titres in hens, vaccinated with a killed bivalent vaccine containing serovars Enteritidis and Typhimurium.

A comparison using the x-OVO Guildhay Typhimurium kit and a commercial swine *Salmonella* (Typhimurium, Infantis and Enteritidis) kit, at a 1:500 dilution with anti-chicken conjugate and substrate added to both kits and read at 550nM showed a 1:0.8 ratio, which was highly significant (Student t-test $P < 0.0001$). While the ratio was not the theorised 1:1 ratio, there was a consistently higher OD observed in the Idexx trivalent kit, which may be attributed to the presence of serogroup C1 antibodies.

The S. Typhimurium ELISA results in experiment 1 and 2 were significantly different (Student t-test $P = 0.001$) between vaccinated and non-vaccinated hens (titre < 460) (Table 4) and results from this trial are very similar to the cited trials and therefore protective levels of antibodies have to be present in vaccinated hens caeca (Clifton-Hadley et al., 2002; Beal et al., 2006; Okamura et al. 2007; Deguchi et al. 2009).

As this was a new vaccine an extremely high co-efficient of variance (127 %) was observed in, the commercial field, vaccinated flocks used in Experiment 1. This may have been due to phase separation of the adjuvant and the killed cells gravity settled to the bottom of the vial. The repeat Experiment (2) the vaccine suspension was vigorously shaken prior to hand vaccination under trial conditions. This resulted in a decreased co-efficient of variation in vaccinated hens to 26 % and a higher percentage positive (36 % >785 titre).

Prior to Experiment 2 (heterologous challenge), all hens were sourced unvaccinated, then manually vaccinated at 12 and 17 weeks to measure the seroconversion rate. The data summarised in Table 5 indicated seroconversion (> 785 titre) which, while slow between the priming dose at 12 weeks of age (8%) and the booster at 17 weeks (18 %), peaked at 43 % by 20 weeks of age. Deguchi *et al.* (2009) also observed significantly high titre values four weeks post vaccination and McMullin *et al.* (1997) indicated that a minimum of 30 % of hens positive serologically by 22 weeks of age was required to confer flock immunity (to colonisation by *S.* Enteritidis using Salenvac vaccine and the Guildhay ELISA for *S.* Enteritidis). In the present study, 43 % of hens were seropositive (titre > 785) after 20 weeks, indicating flock immunity.

The Guildhay method requires an initial serum dilution of 1:500 whereas similar work with killed Typhimurium used initial serum dilution of 1:250, 1:400 and 1:400 respectively (Beal *et al.*, 2005; Withanage *et al.*, 2005; Bailey *et al.*, 2007a). The higher initial dilution rate may have greatly reduced the number of ELISA positives in this trial.

In the vaccinated flock challenged with *S.* Typhimurium (Table 2a) the strain was recovered from the caeca of only one hen (*n* = 11) compared to six hens in the non-vaccinated group (*n* = 12). The vaccination of hens against *Salmonella* may aid in the decrease of *Salmonella* numbers house from the caeca, and thus decrease shell contamination, which may, in turn, prevent chick colonisation during hatching.

The amount of IgY transferred to the egg yolk is proportional to maternal serum IgG concentrations (Loeken and Roth, 1983; Al-Natour *et al.*, 2004; Hamal *et al.*, 2006). The transfer of circulating anti-*Salmonellae* antibody (IgG) into egg yolk (IgY), was demonstrated by Chalghoumi *et al.* (2008) and Bailey *et al.* (2007a), after vaccination with serovars Enteritidis and Typhimurium.

Therefore, it is expected that antibodies against the three vaccine strains used in this study would have been transferred to the yolk. In the present study, the detection of serovar Typhimurium antibodies (titre > 785) were measured in 16 % of egg yolks, and 52 % contained no Typhimurium antibody (titre < 460). In addition, 32 % of yolks gave titres that were in the 'suspect' range for the ELISA (460 to 784).

The data suggest maternal transfer of IgY of Typhimurium antibodies from vaccinated hens. The level of detection in this study was 16 % (> 760) or 48 % (> 460) and was similar to the 30 % achieved in a previous study (Hamal *et al.*, 2006).

The variable level of titre measured in egg yolk may be an artefact from initial dilution (1:500), as discussed previously, especially when compared to the study of Hassan and Curtiss (1996) egg yolk dilution of 1:100. The titres of anti-*Salmonella* antibody in this study ranged from 538-1784 and 145-1890 in serum and yolk respectively (at the 95th percentile), indicating that sero-conversion was not uniform among hens.

Therefore, low level of circulating antibody would likely result in transfer of a low level of antibody into the egg yolk. Also, as the eggs were sampled from pens and not individual hens, only broad comparisons could be made. The presence of anti-Typhimurium IgY in egg yolk from vaccinated hens was not observed as increased anti-Typhimurium IgG in their serum of their progeny.

The final experiment (Experiment 3) measured the effects of maternal anti-*Salmonella* antibodies in day-old chicks, challenged with different doses of *S.* Typhimurium. Reduction in colonisation by S. Typhimurium is problematic, even in vaccinated chicks. While good reduction has been observed when chicks are challenged with 10^4 cfu, reduction is poor at higher challenge doses (10^6 and 10^8 cfu; Nisbet et al., 2000; Clifton-Hadley et al., 2002).

Methner et al. (1999) showed a 0.5-1.5 $\log_{10}$ cfu reduction in carriage in vaccinated chicks challenged with variable populations of S. Typhimurium. Bailey, et al. (2007b) showed a 0.5 $\log_{10}$ reduction in day old progeny from vaccinated hens, challenged with 10^7 cfu, which may be due to maternal IgY.

In the present experiment, in vaccinated chicks, a 25 % clearance from caeca was observed in chicks challenged with 10^4 cfu, while no reduction was evident using a 10^8 cfu challenge (Table 6), indicating a limit to protection by maternal antibodies.

The literature showed conflicting results with respect to maternal antibody protection against Salmonella colonisation; Hassan and Curtiss (1996) showed protection whereas Bailey et al. (2007b) showed no protection. There is a need for more experimental work in this area using standardised ELISA, challenge and measurement techniques.

Regardless of challenge dose, Nisbet et al. (2000) showed a 2 $\log_{10}$ reduction in Salmonella establishing in the caeca whereas Clifton-Hadley et al. (2002) suggested that high challenge doses of highly invasive and virulent S. Typhimurium may overcome those components of the immune system associated with the systemic phase of infection. Therefore, the 10^4 cfu challenge is the minimum dosage that will not overwhelm the immune system and still results in colonisation of non-vaccinated controls.

Based on the results of this study, vaccination of all breeder flocks by several Australian commercial broiler growers was introduced as an adjunct to other Salmonella control measures. The effectiveness of this vaccine in the field was tested using routine monitoring data, for blood and drag swabs for weeks 6, 14, 18, 22, 33, 43, 53 for the year prior (2003) to vaccination, during vaccination (2004) and post vaccination (2005). The results showed an overall decrease in flock prevalence from 51% (2003) to 40 % and 41 % (2004 and 2005 respectively) which, while not statistically significant (χ^2 P = >0.05), was an important reduction. The comprehensive risk assessment produced by the FAO/WHO (2002) stated that "there was a one to one relationship (assuming that everything else remains constant) between percentage change in prevalence to expected risk of illness". The example cited by the FAO/WHO (2002) stated that if prevalence reduced from 20 % to 10 %, this would result in a similar 50 % reduction of risk in illness per serving.

The reduction in flock prevalence was noticed across all flock ages (Figure 3) with a reduction in chick (<12 woa) prevalence, which may be attributed to maternal tri-vaccine antibody (Hassan and Curtiss, 1996). The increase in Salmonella prevalence during weeks 13 to 22 may be due to the hen sexual maturity, which may be associated with a situation akin to the peri-parturient relaxation of resistance (Kelly, 1973) seen in mammals.

This condition is contributory to susceptibility of sheep to parasite infection after birth and has been attributed to the release of prolactin (Kelly, 1973). Hens also produce prolactin, initiating broodiness, which may limit early egg production and may increase their susceptibility to Salmonella colonisation (Leboucher et al., 1990; Talbot et al., 1991; March, et al., 1994; Berry, 2003).

The increase in flock prevalence of *Salmonellae* in that age bracket (13 to 22 weeks) may also be due to the introduction of males from other rearing houses. These males were not vaccinated and may have been a source of *Salmonella*.

There was a decrease in *Salmonella* prevalence post-vaccination, (2004-2005) across all age groups, for serogroups B and C but not for serogorup E (Figure 3 and Table 7). The flock horizontal data are very similar to the observations noticed in the trial (Table 2) and support the hypothesis that serogroup E is transitory and does not persistently colonise the avian intestinal tract.

There was also a change in *Salmonella* serovar profile (Table 7), with 13 serovars in 2003 accounting for 97 % of all flock isolates. When these same serovars were tracked through the vaccination years (2004 to 2005) they accounted for only 88.6 % and 71.9 % respectively. The biggest change in prevalence was observed in the non-vaccine serovars Senftenberg (serogroup E4) from 0.6 % in 2003 to 4.7 % in 2005 and subsp. 1 ser rough:r:1,5 from 0.6 % to 14.6 % for the same time period (Figure 3 and Table 7).

The serovar changes observed may have been introduction through feed, litter, vermin, and equipment (FAO/WHO, 2002; Arsenault *et al.*, 2007). The presence of this rough:r:1, 5 serovar has identical flagellar antigens to Infantis, suggesting these isolates are variants of that serovar. The term "rough" means that this *Salmonella* is devoid of somatic antigens (lipopolysaccharide). However, its flagellar antigenic structure is very similar to that of serovar Infantis (Davos, personal communication).

According to the complete Le Minor-White-Kauffman scheme there are 11 other serovars that have the same flagellar structure: Bradford (serogroup [B]); Czernyring [054]; Abertbanju [V]; Lubumbashi [S]; Hindmarsh $[C_2-C_3]$; Linde [P]; Jamaica $[D_1]$; Ughelli $[E_1]$; Senegal [F]; Tennenlohe [K] and Gege [N] (Grimont and Weill, 2007). However, none of these serovars have been isolated in Australia (Davos, 2008) thus supporting the initial hypothesis that this rough strain is a variant of Infantis.

The flock seroconversion of the Typhimurium component was measured and reported in Figure 4. The annual (2004 and 2005) flock positive (> 785) titre rates (flock immunity) were 28 % and 27 % respectively, with peak rates recorded between 13 and 22 woa. These figures were less than the trial figure of 43 % at 20 woa and slightly lower than the 30 % seroconversion described by McMullin *et al.* (1997). This may be due to difficultly in hand vaccinating thousands of hens per day (shown by the wide 95 % confidence range) and the stress placed upon the hen which may act as an immunosuppressant and compromise seroconversion (Shini *et al.*, 2010).

The data summarised in Figure 5 suggest that a high average titre (> 1000 mean) may aid in reduced shedding of serovar Typhimurium within the flock. This may be due to the increased production of IgY and IgA within the gut of the hen (Bailey *et al.*, 2007a). The mean titre of 1000 was chosen as its S/P ratio was 0.35 which was 0.1 larger than the cut-off positive S/P ratio of 0.25. The cut off age of 25 weeks was chosen as that is approximately the age when the hens are in full fertile egg production.

An interesting observation was seen on a day old to death (Aug to Jun) farm with six houses (flocks) on which all were colonised with Typhimurium prior to vaccination. The flock in the fourth house had a high mean titre (2566 and 75 % flock immunity) and was clear of

Typhimurium colonisation at 23, 33 and 43 woa (no 53 woa swabs taken) whereas the adjoining flocks in houses three and five, with low titres (266 and 786 respectively), were still positive for Typhimurium after 33 and 43 weeks.

In another State, a four-house day old to death (Oct to Sep) complex had the first house with a high mean titre (1246) and was negative for Typhimurium after 25 weeks whereas the adjoining three houses with low titres (426, 828 and 533) were positive after 25 weeks. More controlled research is required to understand the factors contributing to clearance.

As these flocks were on the same farm and the adjoining houses were positive, then it may be safe to assume that the feed and water were *Salmonella* free and biosecurity prevented cross-contamination. However, both of these farms had relatively young flocks over the Australian summer (Nov to Feb). These poultry farms have curtained sides, and the presence of flies is very high. Flies are known carriers of *Salmonella* (Wales *et al.*, 2010) and could continuously reinfect the flocks, therefore limiting the effectiveness of biosercurity in preventing *Salmonella* colonisation and spread.

Another possible explanation is that the isolation of *Salmonella* using a drag swab may favour the dominant serovar within the flock and the sub-dominant serovars may be masked or have a lower probability of detection. However, this error should be consistent throughout the farm and a reduced number of Typhimurium would also mean a reduced risk as mentioned previously.

The two most consistent observations are that both these flocks, in different locations and States, had high (> 1000) Typhimurium titres; as Occam's (or Ockham's) Razor states, "with all things being equal the simplest solution is often the correct one" (Anonymous, 2010). Therefore high titre and flock immunity is very important in clearance of Typhimurium.

One of the problems associated with horizontal field trials is the absence of contemporary controls. Therefore, assumptions are made that all growing conditions, feed, water supply and management are the same (constant variable), with vaccination (controlled variable) and *Salmonella* exposure (uncontrolled variable) the measured variables. Unfortunately, when using commercial broiler flocks, in real world situations, as the time frame increases so too does the possibility that flock management practices may change due to flock illness or economic necessities thus altering crucial parameters. Therefore, any conclusions made from historical data need to be guarded and very general.

In conclusion, the vaccination of Cobb™ breeders with an autologous vaccine demonstrated statistically significant reduction in serogroup B and C colonisation following challenge at 10^7 cfu per bird. Seroconversion of the vaccinated hens, as well as maternal transfer of antibodies to the egg yolk, was shown. Challenge (10^4 cfu per bird) trials with day-old chicks demonstrated a significant difference in colonisation of progeny from vaccinated versus non-vaccinated parents. The horizontal analysis over a three year period (2003 to 2005) showed that the number of colonised flocks decreased by 10 % following vaccination and, if the vaccine titre mean was greater than 1000 the likelihood of persistent colonisation was reduced by 17 times.

4.2 Prophylaxis trial

It is well documented that young (< 3 woa) chicks lack a mature gut flora and immune system and therefore are more susceptible to *Salmonella* colonisation (Nurmi and Ratala

1973; Martin *et al.*, 2000;). The transfer of maternal antibodies from the yolk to the chick may prevent colonisation of chicks by *Salmonella* (Hassan *et al.*, 1996).

The data summarised in Table 9 shows that maternal antibodies for *Salmonella* Typhimurium are transferred from hen serum to the egg yolk and these antibodies can be chemically extracted (Staak *et al.*, 2001). However, there is a large variation in these data (15 % to 80 %) in terms of the number of eggs that were positive for *S*. Typhimurium antibodies. This may be due to the level of seroconversion and the subsequent production of circulating antibody within the donor hen. Hamal *et al.* (2006) showed a relationship between plasma antibody levels and transfer into the yolk. These authors also noticed a farm difference in the serum antibody levels to the same vaccine, though, the serum to egg transfer rate was similar in both farms.

The data in Table 9 shows that hens from farm 3, with a high flock mean Typhimurium serum (ELISA titre of 2104), also had 80 % of their yolks positive. These yolks, when chemically extracted, had a high crude extract mean anti-Typhimurium antibody titre (ELISA titre of 6900). The extract titres were higher than the initial serum level and similar observations have been made in numerous studies (Rose *et al.*, 1974; Kariyawasam *et al.*, 2004; Malik *et al.*, 2006; Yegani and Korver, 2010).

The farm (1) that had the lowest levels of Typhimurium IgY was further investigated by performing repeats (n = 8) of each individual yolk tested (Figure 8). These data (Figure 8) showed that there was a lot of variation within the egg yolk as signified by the large 95% confidence limits error bars. This may be due to either mechanical error, such as difficultly in accurately pipetting 100 µl of highly viscous egg yolk, or that the amount of *S*. Typhimurium IgY present was so low that it reduced the probability of detection. An alternative method would be to weigh approximately 100 µg of egg yolk instead of pipetting.

The yields of crude IgY which was chemically extracted (Staak *et al.*, 2001) from egg yolks was higher (26 to 32 mg/g) than expected (20 to 25 mg/g) from another study employing the same method (Rose and Orlans, 1981). This higher yield may be due to the extraction of some low molecular weight proteins. This observation could be further investigated through electrophoretic analysis of the extracts to determine the size of the impurities.

In the previous Section (4.1) it was shown that maternal antibody to *S*. Typhimurium can confer some protection to chicks challenged with low levels (10^4) of *S*. Typhimurium. This led to the hypothesis that feeding day old chicks rations supplemented with anti-Typhimurium IgY may provide passive immunity. This hypothesis was based upon the review by Schade *et al.* (2005), which described many examples of transfer of antibody against specific pathogens from hyperimmunized hens into the egg. In recent years, there have been many trials that tested the efficacy of such antibodies against *Salmonella in vitro* (Lee *et al.*, 2002; Chalghoumi *et al.*, 2009a) and *in vivo* (Gurtler *et al.*, 2004; Rahimi *et al.*, 2007; Chalghoumi *et al.*, 2009b).

In the present trial the effects of IgY supplemented feeds on the growth rate, as well as protection from colonisation upon challenge with 10^4 and 10^5 cfu of *S*. Typhimurium, were evaluated; the challenge dose was based upon populations likely to be encountered naturally. The initial part of the trial was to observe and measure whether the chicks would consume the treated feeds and the effects of these on weight gain.

One of the concerns was the salt content in the crude yolk extract may make the feed unpalatable or that other antibodies in the crude yolk extract may interact adversely within the gastrointestinal tract thus affecting feed conversion. The results for this aspect of the trial are summarised in Figure 9 and show that there was no significant (Student t-test $P = 0.56$) difference in weight gain between the treated and control chicks, as shown by Kassaify and Mine (2004) in field trials.

Chicks consumed a standard dose of 1.42 g antibody extract/chick/day. This dosage was calculated from Anonymous (2007) and was based upon 3 % (Gurtler et $al.$, 2004) of daily intake at 11 days (trial midpoint) of age.

The titre of the Typhimurium IgY in the treated feeds was measured daily (Figure 10) and showed that there was a high degree of variation in dosage. The feed treated with dried anti-Typhimurium egg yolk (dT-IgY), resuspended in crude S. Typhimurium containing IgY extract, and had the most consistent titre. The dried feed supplemented with 3 % dried egg yolk was the most variable. This may be due to the difficulty of evenly distributing the dried egg yolk throughout the ration; the titres in reconstituted (wet) feeds were much more consistent.

Finally, after 15 days of age, the chicks were euthanized and their caeca removed for enumeration of S. Typhimurium (Figure 11). The results showed, regardless of the treatment and challenge level, that there was no significant difference (Student t-test $P = 0.65$) between treatment and control groups.

The work performed by Kassaify and Mine (2004) measured the effects of egg yolk powder at 2.5 %, 5 %, 7.5 % and 10 % w : w and showed that $Salmonella$ Typhimurium colonisation in poultry was eliminated at only the 10 % w : w concentration. The authors suggested that egg yolk containing anti-infection and adhesive factors, which may agglutinate the pathogen, stimulates the immune system or factors compete for adhesion sites (Kassaify and Mine, 2004).

Based on this result the concentration of egg yolk in the current trial, at 3 %, may have been too low and an insufficient amount of egg yolk containing anti-Typhimurium IgY reached the caeca. The former is supported by the work of Chalghoumi et $al.$ (2009b) who also showed no significant caecal reduction in four-day-old chicks challenged with 10^6 cfu Typhimurium or Enteritidis fed 1 to 5 % (w:w) freeze dried egg yolk containing anti-$Salmonella$ IgY.

The 3 % dose rate was based on the work of Gurtler et $al.$ (2004), which demonstrated a significant reduction in S. Enteritidis contamination of eggs from hens that were challenged with high levels of Enteritidis (10^8 and 10^9). This challenge rate was not used in the present study as it was not considered to represent a natural infective dose.

An alternate delivery method, IgY administered for 28 days in drinking water, led to a reduction in caecal colonisation of Enteritidis, administered at 10^6 cfu, to 0.27 $\log_{10}$ compared to controls, at 3.98 $\log_{10}$/g (Rahimi et $al.$, (2007). This method may be preferable to adding IgY directly into feed, as the water lines can be set to dose at a specific concentration. In standard Australian veterinary practice antimicrobials are added into water to treat disease and in feed for prophylactic treatment.

The initial concentration of egg yolk containing IgY may be critical as these orally-administered antibodies, like any other protein molecule, are susceptible to denaturation by the acidic pH of the proventriculus and gizzard and degradation by proteases (Yegani and Korver, 2010). However, there are reports in the literature that suggest that the IgY F_{ab} fragments maintain their ability to bind antigen, even after exposure to pepsin and trypsin and at pH 4.0 (Shimizu et al., 1988; Carlander et al., 2000; Gurtler et al., 2004). Therefore, there may be a percentage of the yolk containing IgY digested with only a limited portion still being active post-digestion.

These antibodies, post-digestion, would decline in immunological activity and concentration as they progressed from the proximal to distal regions of the intestine, due to viscosity and passage rate, but remain detectable in the caecum and therefore reducing the probability of antibody encountering and binding the antigen (Reilly et al., 1997; Wilkie et al., 2006; Yegani and Korver, 2010). These factors may influence the ability of IgY to prevent colonisation by specific pathogens in the lower parts of the intestinal tract. It may be possible to develop a protease-resistant oral dosage form of IgY in order to increase the fraction of immuno-reactive antibody delivered locally in the gastrointestinal tract (Reilly et al., 1997).

The easiest method to ensure that effective concentration of IgY reaches the caeca is to ensure that the initial concentration of dried egg yolk is high enough to deliver a dosage of antibody inhibitory toward Salmonella into the caeca. The literature reviewed currently suggested that 10 % (w : w) would be the optimum dosage.

In conclusion, this study demonstrated the successful chemical extraction of Typhimurium IgY transferred from serum to egg yolk for three different flocks. However, when yolk containing Typhimurium IgY was added to dry chick feed at 3 % (w:w) and fed prophylactically to day-old chicks there was no protection observed when these chicks were challenged with S. Typhimurium (10^5 cfu/mL). This was possibly due to the feed dosage being too low or an inability of the orally administered antibodies to reach the site of activity required.

5. Conclusion

The autologus Salmonella tri-vaccine was shown to convey protection from colonisation of autologous and heterologous Salmonella challenge in the hen and to convey maternal antibodies to progeny. These maternal antibodies helped protect day-old chicks from colonisation after low-level challenge with Salmonella Typhimurium (at an age of 14 days). However, when these maternal antibodies were extracted from vaccinated hens' eggs and fed to non-vaccinated day old chicks for prophylaxis there was no difference in caecal colonisation rates between treated and untreated groups.

Reduction of the prevalence of Salmonella in flocks will benefit society by a potential reduction in human foodborne illness, as highlighted by a FAO/WHO (2002) risk assessment, which showed that for every 50 % reduction in prevalence there is also a 50 % reduction in risk.

6. References

ABARE (Australian Bureau of Agricultural and Resource Economics) (2008), *Australian Commodity Statistics*, Canberra, ACT. Australia.

Al-Natour, M. Q., Ward, L. A., Saif, Y. M., Stewart, B. and Keck L. D. (2004). Effect of different levels of maternally derived antibodies on protection against infectious bursal disease virus. *Avian Dis.* 48, 177–182.

Animal Research Act 1985. New South Wales Consolidated Regulations, Australia. www.austlii.edu.au/au/legis/nsw/consol_act/ara1985134 {accessed 26/5/07}.

Animal Research Regulation 2005. New South Wales Consolidated Regulations, Australia. www. austlii.edu.au/au/legis/nsw/consol_reg/arr2005225 {accessed 26/5/07}.

Anonymous (1999). Submission to the inquiry into primary producer access to gene technology. *Australian Academy of Science.* www.science.org.au {accessed 18/1/11}.

Anonymous (2004). The use of vaccines for the control of *Salmonella* in poultry. *The EFSA Journal* 114, 1-74.

Anonymous (2007). Cobb 500 Broiler Growth &and Nutrition Supplement Metric Version. Cobb-Vantress Inc Arkansas.

Anonymous (2010). www.Merrian-Webster.com/dictionary/occam'srazor {accessed 23/3/10}

Arsenault, J., Letellier, A., Quessy, S., Normand, V. and Boulianne, M. (2007). Prevalence and risk factors for *Salmonella* spp. and *Campylobacter* spp. caecal colonisation in broiler chicken and turkey flocks slaughtered in Quebec, Canada. *Prev. Vet. Med.* 81, 250-264.

Arzey, G., (2005). Guidelines - Joint NSW/Victoria *Salmonella* Enteritidis monitoring and accreditation program. Elizabeth Macarthur Agricultural Institute, Menangle NSW Department of Primary Industries, Australia.

AS/NZS 5013.11.1 (Australian/New Zealand Standard) (2004). Microbiology of food and animal feeding stuffs-Preparation of test samples, initial suspension and decimal dilutions for microbiological examination-General rules for the preparation of the initial suspension and decimal dilutions. Standards Australia/Standards New Zealand.

Babu, U., Scott, M., Myers, M.J., Okamura, M., Gaines, D., Yancy, H.F., Lillehoj, H., Heckert R.A. and Raybourne R.B. (2003). Effects of live attenuated and killed *Salmonella* vaccine on T-lymphocyte mediated immunity in laying hen. *Vet Immunol Immunopathol*, 91, 39-44.

Bailey, J.S., Rolón, A., Hofacre, C.L., Holt, P.S., Wilson, J.L., Cosby, D.E., Richardson, L.J. and Cox, N.A. (2007b). Resistance to challenge of breeders and their progeny with and without competitive exclusion treatment to Salmonella vaccination programs in broiler breeders. *Internat. J. Poult. Sci* 6, 386-392.

Bailey, J.S., Rolón, A., Holt, P.S., Hofacre, C.L., Wilson, J.L., Cosby, D.E., Richardson, L.J. and Cox, N.A. (2007a) Humoral and mucosal-humoral immune response to a *Salmonella* vaccination program in broiler breeders. *Internat. J. Poult. Sci* 6, 172-181.

Barrow, P.A. (2007). *Salmonella* infections: Immune and non-immune protection with vaccines. *Avian Pathol.* 36, 1-13.

Barrow, P.A., Huggins, M.B. and Lovell, M.A., (1994). Host specificity of *Salmonella* infection in chickens and mice is expressed in vivo primarily at the level of the reticuloendothelial system. *Infect. and Immun.* 62, 4602-4610.

Beal, R.K. and Smith, A.L. (2007). Antibody response to *Salmonella*: its induction and role in protection against avian enteric salmonellosis. *Exp. Rev. Anti-infect. Ther.*, 5, 873-881.

Beal, R.K., Powers, C., Wigley, P., Barrow, P.A., Kaiser, P. and Smith, A.L. (2005). A strong antigen specific T-cell response is associated with age and genetically dependent resistance to avian enteric salmonellosis. *Infect. Immun.* Nov, 7509-7516.

Beal, R.K., Wigley, P., Powers, C., Barrow, P.A., and Smith, A.L. (2006). Cross-reactive cellular and humoral immune responses to *Salmonella* enterica serovars Typhimurium and Enteritidis are associated with protection to heterologous re-challenge. *Vet Immunol Immunopathol* 15, 84-93.

Berry, W.D. (2003). The physiology of induced molting. *Poult. Sci.* 82, 971-980.

Braden, C.R. (2006). *Salmonella enterica,* serovar Enteritidis and eggs: a national epidemic in the United States. *Clin. Infect. Dis,* 43, 512-517.

Buzby, J.C. and Farah, H.A. (2006). Chicken consumption continues long run rise. *Amber Waves* 4, 5.

Byrd, J.A., Anderson R.C., Callaway T.R., Moore R.W., Knape K.D., Kubena L.F., Ziprin R.L. and Nisbet D.J. (2003). Effect of experimental chlorate product administration in the drinking water on *Salmonella* Typhimurium contamination of broilers. *Poult. Sci.* 82, 1403-1406.

Callaway, T.R., Edrington, T.S., Anderson, R.C., Byrd J.A. and Nisbet, D.J. (2008). Gastrointestinal microbial ecology and the safety of our food supply as related to *Salmonella. J. Anim. Sci.* 86, E163-E172.

Carlander, D., Kollberg, H., Wejaker, P. E., and Larsson, A. (2000). Peroral immunotherapy with yolk antibodies for the prevention and treatment of enteric infections. *Immuno. Rrev.,* 21, 1-6. Review.

Chalghoumi, R., Marcq, C., Thewis, A., Portetelle, D. and Beckers, Y. (2009b). Effects of feed supplemented with specific hen egg yolk antibody (immunoglobin Y) on *Salmonella* species caecal colonization and growth performances of challenged broiler chickens. *Poult. Sci.* 88, 2081-2092.

Chalghoumi, R., Thewis, A., Beckers, Y., Marcq, C., Portetelle, D. and Schneider, Y.-J. (2009a). Adhesion and growth inhibitory effect of chicken egg yolk antibody (IgY) on *Salmonella* enterica serovars Enteritidis and Typhimurium in vitro. *Foodborne Pathog. Dis.* 6, 593–604.

Chalghoumi, R., Thewis, A., Portetelle, D. and Beckers Y. (2008). Production of hen egg yolk immunoglobulins simultaneously directed against *Salmonella* Enteritidis and *Salmonella* Typhimurium in the same egg yolk. *Poult. Sci.* 87, 32-40.

Chambers, J.R. and Lu, X. (2002). Probiotics and maternal vaccination for *Salmonella* control in broiler chickens. *J. Appl. Poult. Res.* 11, 320-327.

Clifton-Hadley, F.A., Breslin, M., Venables, L.M., Sprigings, K.A., Cooles, S.W., Houghton, S. and Woodward, M.J. (2002). A laboratory study of an inactivated bivalent iron restricted *Salmonella* enterica serovars Enteritidis and Typhimurium dual vaccine against Typhimurium challenge in chickens. *Vet. Microbiol.* 89, 167–179.

Codex Alimentarius Commission (2004). Discussion paper on risk management strategies for *Salmonella* spp. in Poultry. Joint FAO/WHO food Standard Program Codex committee on food hygiene thirty-sixth session, Washinton DC, United States of America, 29 March – 3 April 2004.

Cox, J.M and Pavic, A (2010). Advances in enteropathogen control in poultry production. *J. Appl. Microbiol.* 109, 25-34.

Davos, D. (2005) (ed). *Australian Salmonella Reference Centre 2004 Annual Report.* Institute of Medical and Veterinary Science. Adelaide. South Australia.

Davos, D. (2007) (ed) *Australian Salmonella Reference centre 2006 annual report.* Institute of Medical and Veterinary Science. Adelaide. South Australia.

Davos, D. (2008) (ed). *Australian Salmonella Reference Centre 2007 Annual Report.* Institute of Medical and Veterinary Science. Adelaide. South Australia.

Deguchi, K., Yokoyama, E., Honda, T., and Mizuno, K. (2009). Efficacy of a novel trivalent inactivated vaccine against the shedding of *Salmonella* in chicken challenge model. *Avain. Dis.* 53, 281-286.

Duchet-Suchaux, M., Mompart, F., Berthelot, F., Beaumont, C., Lechopier, P., and Pardon, P. (1997). Differences in frequency, level, and duration of caecal carriage between four outbred chicken lines infected orally with *Salmonella* Enteritidis. *Avain Dis.,* 41, 559-567.

EFSA (European Food Safety Authority) (2007). The Community summary report on trends and sources of zoonoses, zoonotic agents, antimicrobial resistance and foodborne outbreaks in the European Union in 2006. *EFSA J.* 130, 1–352.

FAO/WHO. (Food and Agriculture Organisation of the United Nation and the World Health Organisation) (2002). Risk assessments of *Salmonella* in eggs and broiler chickens. In *Microbiological risk assessment series 2.* World Health Organization / Food and Agriculture Organization of the United Nations. ftp://ftp.fao.org/docrep/fao/005/y4392e/y4392e00.pdf {accessed 12/01/09}.

Foley, S.L. and Lynn, A.M. (2008). Food animal-associated *Salmonella* challenges: pathogenicity and antimicrobial resistance. *J. Anim Sci.* 86, E173-187.

Franco, D.A. (2005). A survey of *Salmonella* serovars and most probable numbers in rendered-animal-protein meals: inferences for animal and human health. *J. Environ. Health,* 67, 18-22.

Fuller R. (1989). Probiotics in man and animals. *J. Appl Bacteriol,* 66, 365-378.

Giannella, R.A. (1996). *Salmonella,* in *Medical Microbiology,* 4th edition, eds Baron, S., Univesity of Texas Press.

Grimont, P.A.D. and Weill F-X. (2007). Antigenic formulae of *Salmonella* serovars. *WHO Collaborating Centre for Reference and Research on Salmonella.* http://www.pasteur.fr/sante/clre/cadrecnr/salmoms-index.html   {accessed 12/11/07}.

Gurtler, M., Methner, U., Kobilke, H. and Fehlhaber, K. (2004). Effect of orally administered egg yolk antibodies on *Salmonella enteritidis* contamination of hen's eggs. *J. Vet. Med. B.* 51, 129-134.

Gurtler, M., Methner, U., Kobilke, H. and Fehlhaber, K. (2004). Effect of orally administered egg yolk antibodies on *Salmonella enteritidis* contamination of hen's eggs. *J. Vet. Med. B.* 51, 129-134.

Hamal, K.R., Burgess, S.C., Pevzner, I.Y. and Erf, G.F. (2006). Maternal Antibody Transfer from Dams to Their Egg Yolks, Egg Whites, and Chicks in Meat Lines of Chickens *Poult Sci* 85, 1364–1372.

Hancock, D.D., Blodgett, D. and Gay, C.C., (1988). The collection and submission of samples for laboratory testing. *Vet. Clin. N. Amer: Food Animal. Pract.* 4, 33-59.

Hassan, J.O. and Curtiss, R. (lll) (1996). Effect of vaccination of hens with an avirulent stain of *Salmonella* Typhimurium on immunity of progeny challenged with wild-type *Salmonella* strains. *Infect. Immun.* 64, 938-944.

ISO 6579:2002. Microbiology of food and animal feeding stuffs – Horizontal method for the detection of *Salmonella* spp. International Organization for Standardization, Geneva, Switzerland.

Jay, S., Davos, D., Dundas, M., Frankish, E., and Lightfoot, D. (2003). *Salmonella*. In *Foodborne microorganisms of public health significance.* Hocking, A.D. (ed). 6th edition, Australian Insitute of Food Science and Technology Incorporated NSW Branch, Food Microbiology Group. Southwood Press Pty Ltd, NSW Aust. 207-267

Kariyawasam, S., Wilkie, B.N. and Gyles, C.L. (2004). Resistance of broiler chickens to *Escherichia coli* respiratory tract infection induced by passively transferred egg-yolk antibodies. *Vet. Microbiol.* 98, 273-284.

Kassaify, Z.G., and Mine, Y. (2004). Nonimmunized egg yolk powder can suppress the colonization of *Salmonella* Typhimurium, *Escherichia coli* O157:H7, and *Campylobacter jejuni* in laying hens. *Poult. Sci.* 83, 1497–1506.

Kelly, J.D. (1973). Immunity and epidemiology of helminthiasis in grazing animals. *New Zeal. Vet. J.* 21, 183-194.

Kinde H., Castellan, D.M., Kerr, D., Campbell, J., Breitmeyer, R. and Ardans, A., (2005). Longitudinal monitoring of two commercial layer flocks and their environments for *Salmonella* enterica Serovar Enteritidis and other *Salmonella*e. *Avian Dis.* 49, 189-194.

Kottom, T.J., Nolan, L.K. and Brown J. (1995). Invasion of Caco-2 cells by *Salmonella* Typhimurium (Copenhagan) isolates from healthy and sick chickens. *Avian Dis* 39, 867-872.

Kowalczyk, Daiss, J., Halpern, J. and Roth, T.F. (1985). Quantitation of maternal–fetal IgG transport in the chicken. *Immuno.* 54, 755–762.

Leboucher, G., Richard-Yris, M.A., Williams, J. and Chadwick, A. (1990). Incubation and maternal behaviour in domestic hens: influence of the presence of chicks on circulating luteinising hormone, prolactin and oestradiol and on behaviour. *Br. Poult. Sci.* 31, 851–862.

Lee, E.N., Sunwoo, H.H., Menninen, K. and Sim, J. S. (2002). In Vitro Studies of Chicken Egg Yolk Antibody (IgY) against *Salmonella* Enteritidis and *Salmonella* Typhimurium. *Poult. Sci.* 81, 632–641.

Lloyd, A.B., Cumming, R.B. and Kent R.D. (1977). Prevention of *Salmonella* Typhimurium infection in poultry by pre-treatment of chickens and poults with intestinal extracts. *Austral. Vet. J.* 53, 82-87.

Loeken, M.R. and Roth T. F. (1983). Analysis of maternal IgG subpopulations which are transported into the chicken oocyte. *Immuno.* 49, 21–28.

Lutful Kabir, S.M. (2010). Avian Colibacillosis and Salmonellosis: A Closer Look at Epidemiology, Pathogenesis, Diagnosis, Control and Public Health Concerns. *Int. J. Environ. Res. Public Health,* 7, 89-114.

Malik, M.W., Ayub, N. and Qureshi, I.Z. (2006). Passive immunization using purified IgYs against infectious bursal disease of chickens in Pakistan. *J. Vet. Sci* 7, 43-46.

Marangon, S. and Busani, L. (2007). The use of vaccination in poultry production. *Rev. Sci. Tech.* 26, 265-274.

March, J.B., Sharp, P.J., Wilson, P.W. and Sang, H.M. (1994). Effect of active immunization against recombinant-derived chicken prolactin fusion protein on the onset of broodiness and photoinduced egg laying in bantam hens. *J. Reprod. Fertil.* 101, 227–233.

Martin, C., Dunlap, E., Caldwell, S. and Barnhart, E., (2000). Drinking water delivery of a defined competitive exclusion culture (pre-empt) in 1 day old broiler chicks. *J. Appl. Poultry Res.* 9, 88-91.

McMullin, P.F., Gooderham, K.R. and Hayes G. (1997). A commercial *Salmonella* Enteritidis ELISA test: results arising for its use in monitoring for infection and response to an inactivated vaccine. *Presented at the World Veterinary Poultry Association Congress, Budapest.* http://www.poultry-health.com/library/wvpasens.htm {accessed 03/12/2003}

Mead, G.C., and Barrow, P.A. (1990). *Salmonella* control in poultry by competitive exclusion or immunisation., *Let. Appl. Microbiol.* 10, 221-227.

Martin, S.W., Meek, A.W., and Willeberg, P. (1988). *Veterinary Epidemiology: Principles and Methods.* Iowa State University Press, Ames. p45.

Methner, U., Barrow, P.A., Berndt, A. and Steinbach, G. (1999). Combination of vaccination and competitive exclusion to prevent *Salmonella* colonisation in chickens: experimental studies. *Intern. J. Food Microbiol.* 49, 35-42.

NHMRC (National Health and Medical Research Council) (2008). *Guidelines to promote the wellness of animals used for scientific purposes. The assessment and alleviation of pain and distress in research animals.* NHMRC Canberra, Australia.

NHMRC/ARC (National Health and Medical Research Council, Australian Research Council and Universities Australia) (2007). *Australian code for the responsible conduct of research.* . NHMRC Canberra, Australia.

Nisbet, D.J., Anderson, R.C., Corrier, D.E., Harvey, R.B. and Stanker, L.H. (2000). Modeling the survivability of *Salmonella* Typhimurium in chicken cecae using an anerobic continous-culture of chicken cecal bacteria. *Microbial. Eco. Health Dis.* 12, 42-47.

Nurmi, E and Ratala, M., (1973). New aspects of *Salmonella* infection in broiler production. *Nat.* 241, 210-211.

Okamura, M., Tachizaki, H., Kudo, T., Kiluchi, S., Suzuki, A., and Nakamura, M. (2007). Comparative evaluation of a bivalent killed *Salmonella* vaccine to prevent egg contamination with *Salmonella enterica* serovars Enteritidis, Typhimurium, and Gallinarum biovar Pullorum, using 4 different challenge models. *Vacc* 25, 4837-4844.

Pavic, A., Groves, P.J., Bailey, G., and Cox, J.M. (2010). A validated miniaturized MPN method, based on ISO 6579:2002, for the enumeration of *Salmonella* from poultry matrices. *J. Appl. Microbiol.,* 109, 25-34

Petrie, A. and Watson, P., (1999). *Statistics for veterinary and animal science.* Blackwell Publishing. Oxford, UK.

Quinn, P.J., Carter, M.E., Markey, B. and Carter G.R. (2000). Enterobacteriaceae in *Clinical Veterinary Microbiology* (pp209-237). Mosby International Limited of Harcourt Publishers Limited, Spain.

Rahimi, S., Shiraz, Z. M., Salehi, T. Z., Torshizi, M. A. K. and Grimes, J. L. (2007). Prevention of *Salmonella* infection in poultry by specific egg-derived antibody. *Int. J. Poult. Sci.* 6, 230–235.

Reilly, R.M., Domingo, R. and Sandhu, J. (1997). Oral delivery of antibodies. Future pharmacokinetic trends. *Clin. Pharmaco.* 32, 313-323.

Rose, M. E. and Orlans, E. (1981). Immunoglobulins in the egg, embryo and young chick. *Dev. Comp. Immuno.* 5, 15-20.

Rose, M.E., Orlans, E. and Buttress, N. (1974). Immunoglobulin classes on the hen's egg: their segregation in yolk and white. *Eur.o J. Immuno* 4, 521-523.

Schade, R. Calzado, E.G., Sarmiento, R., Chacana, P.A., Porankiewicz-Asplund, J., and Terzolo, H.R. (2005). Chicken egg yolk antibodies (IgY-technology): a review of progress in production and use in research and human and veterinary medicine. *Altern. Lab. Anim.* 33, 129-154.

Shimizu, M., Fitzsimmons, R. and Nakai, S. (1988). Anti-E. coli Immunoglobulin Y isolated from egg yolk of immunized chickens as a potential food ingredient. *J. Food Sci.* 53, 1360–1366.

Shini, S., Huff, G.R., Shini, A. and Kaiser, P. (2010). Exploration of cytokine and chemokine gene profiles in chicken peripheral leukocytes. *Poult. Sci.* 89, 841-851.

Smith T. (2003). A focus on *Salmonella*, Food Science Research Information Office, www.nal.usda.gov/fsrio/research/fsheet/fsheet10.htm#class.

Smith, H.W (1956) The use of live vaccines in experimental *Salmonella gallinarum* infection in chickens with observations on their interference effect. *J. Hyg. Camb.* 54, 419-432.

Staak, C., Schwarzkopf, C., Behn, I., Hommel, U., Hlinak, A., Schade, R. and Erhard, M. (2001). Isolation if IgY from yolk. In *Chicken egg yolk antibodies, production and application, IgY-technology*. Schade, R., Behn, I., Erhard, M., Hlinak, A., Staak, C. (Eds). Springer-Verlag Berlin, Heidelberg, New York. 78-84.

StatSoft (2001) *Statistica* Version 6. www.statsoft.com.

Talbot, R.T., Hanks, M.C., Sterling, R.J., Sang, H.M. and Sharp, P.J. (1991). Pituitary prolactin messenger ribonucleic acid levels in incubating and laying hens: effects of manipulating plasma levels of vasoactive intestinal polypeptide. *Endocrinol.* 129, 496-502.

Thrusfield M (2005). Observation. in Veterinary Epidemiology 3rd edition (pp 279) Blackwell Publishing.

U.S-FDA (United.States of America Food and Drug Administration) (2006) Most probable number from serial dilutions. In *Bacterial Analytical Manual online* Appendix 2. United States Food and Drug Administration, Center of Food Safety and Applied Nutrition, Department of Health and Human Services. http://www.cfsan.fda.gov/~ebam/bam-a2.html [accessed 3/3/08].

Van den Bosch, G. (2003). Vaccination versus treatment: how Europe is tackling the eradication of *Salmonella*. *Asian Poult. Mag.* July, 1-3.

Van Immerseel, F., Methner, U., Rychlik, I., Nagy, B., Velge, P., Martin, G., Foster, N., Ducatelle, R. and Barrow P.A. (2005). Vaccination and early protection against non-host-specific *Salmonella* serovars in poultry: exploitation of innate immunity and microbial activity. *Epidemiol. Infect.* 133, 959-978.

Vanderplas, S., Dubois Dauphin, R., Beckers, Y., Thonart, P., and Thewis, A. (2010). *Salmonella* in chicken: Current and developing strategies to reduce contamination at farm level. *J. Food Prot.* 73. 774-785.

Volkova, V.V., Bailey, R.H. and Wills, R.W., (2009). *Salmonella* in broiler litter and properties of soil at farm location. PLoS ONE 4(7): e6403. doi:10.1371/journal.pone.0006403.

Wales, A. D., Carrique-Mas, J. J., Rankin, M., Bell, B., Thind, B. B. and Davies, R. H. (2010), Review of the carriage of zoonotic bacteria by arthropods, with special reference to *Salmonella* in mites, flies and litter beetles. *Zoonoses Public Health*, 57, 299–314.

Wilkie, D.C., Van Kessel, A.G., Dumonnceaux, T. J. and Drew, M. D. (2006). The effect of hen egg antibodies on *Clostridium perfringens* colonization in the gastrointestinal tract of broiler chickens. *Prevent. Vet. Med.* 74, 279-292.

Withanage, G.S.K., Wigley, P., Kaiser, P., Mastroeni, P., Brooks, H., Powers, C., Beal, R., Barrow, P., Maskell, D., and McConnell, I. (2005). Cytokine and chemokine responses associated with clearance of a primary *Salmonella enterica* serovar Typhimurium infection in the chicken and in protective immunity to rechallenge. *Infect. Immun.* Aug, 5173-5182.

Woodward, M.J., Gettinby, G., Breslin, M.F., Corkish, J.D. and Houghton, S. (2002). The efficacy of Salenvac, a *Salmonella enterica* subsp. Enterica serovar Enteritidis iron-restricted bacterin vaccine, in laying chickens. *Avian Pathol.* 31, 383-392.

x-OVO (Guildhay) Flockscreen™ *Salmonella* Typhimurium (St) antibody kit Instruction for use (v4). Inverkeithing, Scotland, UK.

Yegani, M. and Korver, D. R. (2010). Application of egg yolk antibodies as replacement for antibiotics in poultry. *World Poult. Sci. J.*, 66, 27-37.

Zhang-Barber, L., Turner, A.K., and Barrow, P.A. (1999). Vaccination for the control of *Salmonella* in poultry. *Vacc.* 17, 2538-2545.

5

Lipoxygenase-Quercetin Interaction: A Kinetic Study Through Biochemical and Spectroscopy Approaches

Veronica Sanda Chedea[1,2], Simona Ioana Vicaş[2], Carmen Socaciu[3],
Tsutomu Nagaya[4], Henry Joseph Oduor Ogola[4,5],
Kazushige Yokota[4], Kohji Nishimura[4] and Mitsuo Jisaka[4*]

1. Introduction

1.1 Lipoxygenase – definition, structure, reaction mechanism and metabolic functions

Lipoxygenases (EC 1.13.11.12, linoleate:oxygen, oxidoreductases, LOXs) which are widely found in plants, fungi, and animals, are a large monomeric protein family with non-heme, non-sulphur, iron cofactor containing dioxygenases that catalyze the oxidation of polyunsaturated fatty acids (PUFA) as substrate with at least one 1Z, 4Z-pentadiene moiety such as linoleic, linolenic and arachidonic acid to yield hydroperoxides (Gardner, 1991).

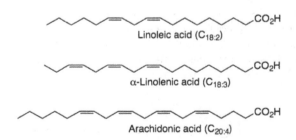

Fig. 1. Lipoxygenase substrates, linoleic, α-linolenic and arachidonic acid.

*1 Laboratory of Animal Biology, National Research Development Institute
for Animal Biology and Nutrition Baloteşti (IBNA), Romania
2 Faculty of Environmental Protection, University of Oradea, Romania
3 Department of Chemistry and Biochemistry, University of Agricultural Sciences and Veterinary Medicine Cluj-Napoca, Romania
4 Department of Life Science and Biotechnology, Faculty of Life and Environmental Science, Shimane University, Japan
5 School of Agriculture, Food Security and Biodiversity, Bondo University College, Kenya

The crystal structures of soybean LOX-1 (Minor et al., 1996), LOX-3 (Skrzypczak-Jankun et al., 1997), rabbit 15S-LOX-1 (Gillmor et al., 1997), and coral 8R-LOX (Oldham et al., 2005) have been elucidated thus far and has helped in understanding the properties of LOXs at the molecular level. Several structures of soybean LOX in complex with its product or inhibitor are also known (Skrzypczak-Jankun et al., 2001; Skrzypczak-Jankun[a] et al., 2003; Skrzypczak-Jankun[b] et al., 2003; Borbulevych et al., 2004; Skrzypczak-Jankun et al., 2004). There are significant differences in size, sequence, and substrate preference between the plant and animal LOXs, but the overall folding and geometry of the nonheme iron-binding site are conserved (Kühn et al., 2005; Skrzypczak-Jankun et al., 2006). All LOXs are folded in a two-domain structure that is composed of a smaller β-barrel domain (N-terminal domain) and a larger α-helical catalytic domain (C-terminal domain) (Choi et al., 2008). The nonheme iron essential for activity is positioned deep in a large cavity that accommodates the substrate (Choi et al., 2008). The regio- and stereospecificities of the various LOX isozymes are believed to be determined by the shape and depth of the cavity as well as the binding orientation of the substrate in the cavity (Borngräber et al., 1999; Kühn, 2000; Coffa et al., 2005).

The initial step of LOX reaction is removal of a hydrogen atom from a methylene unit between double bonds in substrate fatty acids (Fig. 2A). The resulting carbon radical is stabilized by electron delocalization through the double bonds. Then, a molecular oxygen is added to the carbon atom at +2 or –2 position from the original radical carbon, forming a peroxy radical as well as a conjugated *trans, cis*-diene chromophore. The peroxy radical is then hydrogenated to form a hydroperoxide. The initial hydrogen removal and the following oxygen addition occur in opposite (or antarafacial) sides related to the plane formed by the 1Z, 4Z-pentadiene unit. In most LOX reactions, particularly those in plants, the resulting hydroperoxy groups are in S-configuration, while one mammalian LOX and some marine invertebrate LOXs produce R-hydroperoxides. Even in the reactions of such "R-LOXs", the antarafacial rule of hydrogen removal and oxygen addition is conserved (Chedea & Jisaka, 2011).

In cases of plant LOXs, including soybean LOXs, the usual substrates are C18-polyunsaturated fatty acids (linoleic and α-linolenic acids), and the products are their 9S- or 13S-hydroperoxides (Fig. 2B). Most plant LOXs react with either one of the regio-specificity, while some with both. Therefore, based on the regio-specificty, plant LOXs are classified into 9-LOXs, 13-LOXs, or 9/13-LOXs (Chedea & Jisaka, 2011).

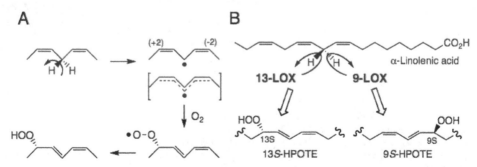

Fig. 2. LOX reaction showing the principal steps of LOX reaction (Panel A), and the actual reactions of plant LOXs and α-linolenic acid (Panel B). HPOTE: hydroperoxyoctadecatrienoic acid (Chedea & Jisaka, 2011).

Theorell et al. (1947) succeeded in crystallizing and characterizing lipoxygenase (LOX) from soybeans and since then among plant LOXs, soybean LOX-1 can be regarded as the mechanistic paradigm for these nonheme iron dioxygenases (Coffa et al., 2005; Minor et al., 1996; Fiorucci et al., 2008).

LOX isoenzymes of soybean seed are 94–97 kDa monomeric proteins with distinct isoelectric points ranging from about 5.7 to 6.4, and can be distinguished by optimum pH, substrate specificity, product formation and stability (Siedow, 1991). LOX-1 is the smallest in size (838 amino acids; 94 kDa), exhibits maximal activity at pH 9.0, and converts linoleic acid preferentially into the 13-hydroperoxide derivative. LOX-2 is characterized by a larger size (865 amino acids; 97 kDa), by a peak of activity at pH 6.8, and forms equal amounts of the 13- and 9-hydroperoxide compounds (Loiseau et al., 2001). LOX-2 oxygenates the esterified unsaturated fatty acid moieties in membranes in contrast to LOX-1 which only uses free fatty acids as substrates (Maccarrone et al., 1994). LOX-3 (857 amino acids; 96.5 kDa) exhibits its maximal activity over a broad pH range centred around pH 7.0 and displays a moderate preference for producing a 9-hydroperoxide product. It is the most active isoenzyme with respect to both carotenoid cooxidation and production of oxodienoic acids (Ramadoss, 1978).

Arachidonic acid metabolism by LOX in platelets was demonstrated in 1974 (Hamberg & Samuelsson, 1974). Rabbit reticulocyte LOX was described for the first time in 1975 (Schewe et al., 1975). There are 6 functional LOXs in humans: 5-LOX; 12/15-LOX (15-LOX-1); platelet-type 12-LOX; 12R-LOX; epidermis-type 15-LOX (15-LOX-2); and epidermis-Alox3. Each of these genes expresses a catalytically active enzyme except epidermis-Alox3, which encodes an enzyme that has hydroperoxidase activity. Thus, the main LOX enzymes with fatty acid oxygenase activity found in humans are 5-LOX, 15-LOX-1, 15-LOX-2, platelet-type 12-LOX and 12R-LOX (Bhattacharya et al., 2009).

In their review concerning the metabolic functions of LOX, Ivanov et al. (2010) present two aspects of LOX biology: (i) 30 years ago, leukotrienes generated by 5-LOX were identified as potent pro-inflammatory mediators (Samuelsson et al., 1980; Jakschik & Lee, 1980). Since then, additional pro-inflammatory products have been discovered (Feltenmark et al., 2008). On the other hand, anti-inflammatory and/or pro-resolving lipids generated by mammalian isoforms have also been identified, including lipoxins (Serhan et al., 1984; Maderna & Gordon, 2009), resolvins (Serhan et al., 2004), protectins (Ariel et al., 2005; Schwab et al., 2007) and maresins (Serhan et al., 2009). Therefore, LOX products play important roles in the development of acute inflammation but they have also been implicated in inflammatory resolution. (ii) Mice deficient in 12R-LOX develop normally during pregnancy but die immediately after birth due to excessive dehydration (Furstenberger et al., 2007; Epp et al., 2007). Although the molecular mechanisms of postpartum mortality are unknown, the enzyme was implicated in the formation of the epidermal water barrier. Genetic polymorphisms of the corresponding human gene have been related to ichthyosis (Eckl et al., 2009), a disease characterized by dry, thickened, scaly or flaky skin (Ivanov et al., 2010).

Available data also suggest that lipoxygenases contribute to *in vivo* metabolism of endobiotics and xenobiotics in mammals (Kulkarni, 2001). Recent reviews describe the role of lipoxygenase in cancer (Bhattacharya et al., 2009; Pidgeon et al., 2007; Moreno, 2009), inflammation (Duroudier et al., 2009; Hersberger, 2010) and vascular biology (Chawengsub et al., 2009; Mochizuki & Kwon, 2008) and for an extensive presentation of the role of eicosanoids in prevention and management of diseases the reader is referred to the review of Szefel et al. (2011).

1.2 Quercetin

Quercetin (3,3′,4′,5,7-pentahydroxyflavone) (Fig. 3) is an important dietary flavonoid, present in different vegetables, fruits, seeds, nuts, tea and red wine (Beecher et al., 1999; Formica, 1995; Hollman & Katan, 1999).

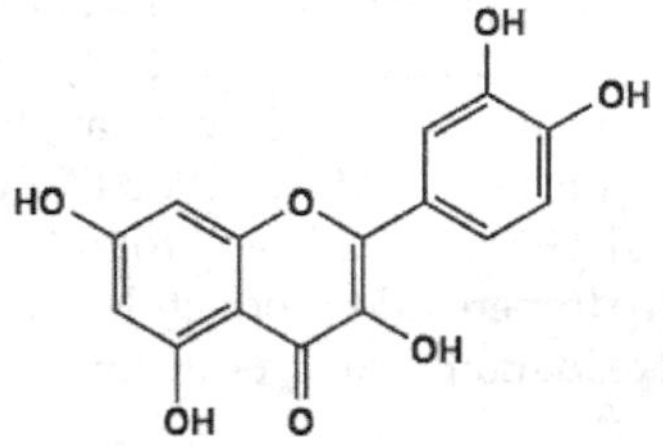

Quercetin

Fig. 3. Chemical structure of quercetin.

Quercetin has been discussed for several decades as a multipotent bioflavonoid with great potential for the prevention and treatment of disease (Bischoff, 2008). Its documented impact on human health includes cardiovascular protection, anticancer, antiviral, anti-inflammatory activities, antiulcer effects and cataract prevention. The study of quercetin as potential chemopreventer is assuming increasing importance considering its involvement in the suppression of many tumor-related processes including oxidative stress, apoptosis, proliferation and metastasis. Quercetin has also received greater attention as pro-apoptotic flavonoid with a specific and almost exclusive activity on tumor cell lines rather than normal, non-transformed cells (Lugli et al., 2009).

Among the biological effects of particular relevance, the antihypertensive effects of quercetin in humans and the improvement of endothelial function should also be emphasized (Bischoff, 2008). Together with its antithrombotic and anti-inflammatory effects, the latter mainly mediated through the inhibition of cytokines and nitric oxide, quercetin is a candidate for preventing obesity-related diseases (Bischoff, 2008). Most exciting findings are that quercetin enhances physical power by yet unclear mechanisms. The anti-infectious and immunomodulatory activities of quercetin might be related to this effect (Bischoff, 2008).

Like other flavonoids, quercetin appears to combine both lipoxygenase-inhibitory activities and free radical-scavenging properties in one agent and thus belongs to a family of very effective natural antioxidants (Sadik et al., 2003). Quercetin is a flavonol that can be easily oxidized in an aqueous environment, and in the presence of iron and hydroxyl free radicals (Borbulevych et al., 2004). More specifically, in an aqueous solution, quercetin is known to be oxidized to the relatively stable, neutral protocatechuate intermediate 2-(3,4-dihydroxybenzoyl)-2,4,6-trihydroxybenzofuran-3(2H)-one (Fig. 5). This protocatechuate derivative is a common oxidized intermediate mediated by various oxidases, such as lipoxygenase, tyrosinase (Kubo et al., 2004), and peroxidase (Awad et al., 2000), as well as by diphenylpicrylhydrazyl (DPPH) and azobisisobutyronitrile (AIBN) (Krishnamachari et al., 2002) and seems to be a key intermediate to understand quercetin's diverse biological activities (Ha et al., 2010).

1.3 LOX inhibition by quercetin

Because of its unique capability of the direct catalysis the enzymatic lipid peroxidation, 15-LOX-1 belongs to the endogenous prooxidants the action of which may be favored under conditions of oxidative stress (Schewe, 2002). Consequently, the inhibition of 15-LOX-1 may contribute to the universal antioxidant activities of dietary flavonoids (Sadik et al., 2003). Flavonoids appear to combine both lipoxygenase-inhibitory activities and free radical-scavenging properties in one agent and thus constitute a family of very effective natural antioxidants (Sadik et al., 2003). The literature data indicate that quercetin represents one of the most potent inhibitors of different LOXs (Schneider & Bucar[a], 2005; Schneider & Bucar[b], 2005).

The inhibition of rabbit 15-LOX-1 and of soybean LOX-1 by quercetin was studied in detail (Sadik et al., 2003). Quercetin modulates the time course of the lipoxygenase reaction in a complex manner by exerting three distinct effects: (i) prolongation of the kinetic lag period, (ii) instant decrease in the initial rate after the lag phase being overcome, (iii) time-dependent inactivation of the enzyme during reaction, but not in the absence of substrate (Schewe & Sies, 2003). Competitive reversible and irreversible inhibition schemes, as well as inhibition via reduction of the enzyme-bound radical intermediate have been considered to explain the activity of polyphenolic compounds (Fiorucci et al., 2008). Moreover, heterogeneity in the interpretation of the experimental results of the inhibition processes, for example concerning kinetic data, prevents converging toward a general way of inhibition (Fiorucci et al., 2008). It may be supposed that the inactivation is due to combined action of quercetin and intermediates of the catalytic cycle on the active site of the enzyme (Sadik et al., 2003). In previous work, Redrejo-Rodriguez et al. (2004) reported by semiempirical studies that the interaction of quercetin with lipoxygenase is related to the spatial adaptation of the flavonoid to the hydrophobic cavity that constitutes the channel of access of substrate to the catalytic site.

Structural analysis reveals that quercetin entrapped within LOX undergoes degradation and the resulting compound has been identified by X-ray analysis as protocatechuic acid (3,4-dihydroxybenzoic acid) positioned near the iron site (Borbulevych et al., 2004).

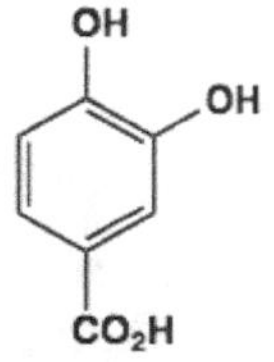

Protocatechuic acid

Fig. 4. Chemical structure of protocatechuic acid, the product of quercetin degradation by soybean LOX-3 as identified by Borbulevych et al. (2004).

The finding that LOX can turn different compounds, like quercetin and epigallocatechin gallate, into simple catechol derivatives (with one aromatic ring only) might be of importance as an additional small piece of a "jigsaw puzzle" in the much bigger picture of drug metabolism. Their interactions with LOX can be more complicated than simply blocking the access to the enzyme's active site (Borbulevych et al., 2004). Ha et al. (2010) have studied the inhibitory activity of protocatechuic acid and of dodecyl protocatechuate on soybean LOX-1 oxidizing linoleic acid. Their results show that the protocatechuate derivative, dodecyl protocatechuate, inhibited the enzymatic peroxidation of linoleic acid as a competitive inhibitor, but its parent compound, protocatechuic acid did not show any activity up to 200 μM. In this way it was shown that the catechol moiety alone is not sufficient to elicit the inhibitory activity and that the hydrophobic dodecyl group is associated with inhibitory activity as it was also reported for dodecyl gallate (Ha & Kubo, 2007; Ha et al., 2010).

Fiorucci et al. (2008) report theoretical investigations concerning three binding modes between quercetin and LOX-3 enzyme. Thus O3, O7, and O4' oxygen atoms have been considered to bind the iron center (Fiorucci et al., 2008). These specific interactions lead then, through electron transfer from the substrate to the cation, to semiquinone forms, and related tautomeric structures that are compatible with an addition of the triplet spin state dioxygen to the flavonol squeleton (Fiorucci et al., 2008). Among the three considered modes of binding, it appears that quercetin should be linked to the metal center via its 3-OH functional group. The most favorable term to the binding free energy is due to electrostatic interactions (Fiorucci et al., 2008).

In their proposed mechanism of oxidative degradation of quercetin by soybean LOX-1, Ha et al. (2010) indicate that quercetin might first be oxidized to the corresponding o-quinone after the abstraction of 2 e- and 2 H+ from the OH groups at C3' and C4' (Jungbluth et al., 2000). The enzymatically oxidized o-quinone might be subsequently isomerized to a p-quinone methide type intermediate, followed by the addition of H_2O at C2, yielding the relatively stable intermediate 2-(3,4-dihydroxybenzoyl)-2,4,6-trihydroxybenzofuran-3(2H)-one (Ha et al., 2010). This enzymatically-generated intermediate appears to be relatively stable, and hence, prolongs the inhibitory activity (Ha et al., 2010). The results indicate that the soybean LOX-1 generated intermediate of quercetin is the same as the DPPH (2,2-diphenyl-β-picrylhydrazyl) generated intermediate, 2-(3,4-dihydroxybenzoyl)-2,4,6-trihydroxybenzofuran-3(2H)-one (Krishnamachari et al., 2002). In addition, the same oxidized intermediate was also characterized as the relatively stable intermediate generated by mushroom tyrosinase (Ha et al., 2010).

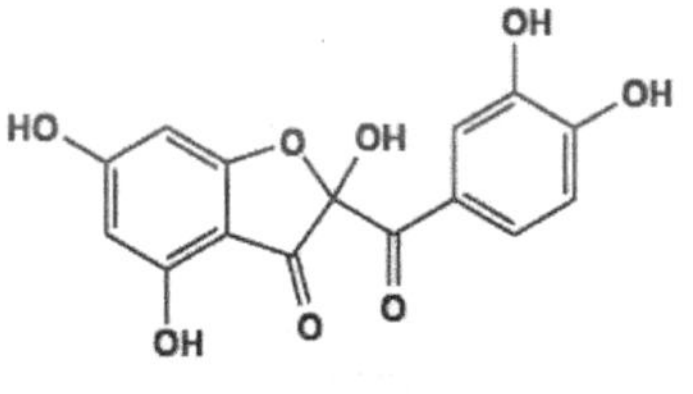

2-(3,4-Dihydroxybenzoyl)-2,4,6-
trihydroxybenzofuran-3(2H)-one

Fig. 5. Product of oxidative degradation of quercetin by soybean LOX-1 as proposed by Ha et al. (2010).

The studies on LOX and quercetin contribute to the understanding of biocatalytic properties of this enzyme and its role in the metabolism of this popular (as a medicinal remedy) flavonol and possibly other, similar compounds (Borbulevych et al., 2004).

1.4 O-quinone formation during quercetin oxidation by LOX

Because flavonoids, such as quercetin, and oxidases are present simultaneously in fruits and vegetables, the generation of quinoid derivatives in biological systems is plausible (Pinto & Macias, 2005). This process is of great relevance from a biological point of view, because the conversion of supposed beneficial antioxidants such as flavonoids to electrophilic prooxidants may constitute a possible toxicological risk (Boersma et al., 2000).

On the other hand, it is to be noted that, in addition to dioxygenase activity, lipoxygenase, possesses a peroxidase activity toward a wide range of compounds (Gardner, 1991). Dioxygenase activity produces the insertion of oxygen into a polyunsaturated fatty acid containing a 1,4-*cis,cis*-pentadiene moiety, producing the corresponding lipid hydroperoxide. In the process, an intermediate peroxyl radical is generated. This compound, or the peroxide, supports the cooxidase activity of lipoxygenase toward a suitable electron donor, which is transformed into a radical (Pinto & Macias, 2005). The hydroperoxidase activity of LOX also can be observed in the presence of hydrogen peroxide instead of lipid hydroperoxide, being related to xenobiotic oxidation processes (Hover & Kulkarni, 2000; Santano[a] et al., 2002; Santano[b] et al., 2002). In addition, it is known that a variety of phenolic compounds and flavonoids with antioxidant properties are inhibitors of lipoxygenase (Prasad et al., 2004).

On the basis of the results obtained, Pinto & Macias (2005) concluded that in the presence of hydrogen peroxide or hydroperoxylinoleic acid, lipoxygenase produces a quinoid product as a result of the enzymatic oxidation (in the presence of hydrogen peroxide) or cooxidation (in the presence of linoleic acid) of quercetin (Pinto & Macias, 2005).

During lipoxygenase catalysis enzyme-bound prooxidant intermediates such as fatty acid peroxyl radical (ROO$^{\bullet}$) are formed (Schewe, 2002). It is tempting to speculate, therefore, that the flavonoids are co-oxidized in this system to a semi-quinone or quinone (with flavonoids containing a catechol B ring) or a phenoxy radical (with noncatechol flavonoids) which in turn may covalently bind to sulfhydryl or amino groups of the lipoxygenase, thus rendering its inhibition irreversible (Sadik et al., 2003). In the case of quercetin and other flavonols, the intermediate formation of corresponding quinone methides (Awad et al., 2001) may be involved (Sadik et al., 2003).

Quercetin is considered an excellent free radical scavenging antioxidant owing to the high number of hydroxyl groups and conjugated π orbitals by which quercetin can donate electrons or hydrogen, and scavenge H_2O_2 and superoxide anion ($\bullet O^{2-}$) (Heijnen et al., 2001). The reaction of quercetin with $\bullet O^{2-}$ leads to the generation of the semiquinone radical and H_2O_2 (Metodiewa et al., 1999). Quercetin also reacts with H_2O_2 in the presence of peroxidases, and thus it decreases H_2O_2 levels and protects cells against H_2O_2 damage; nevertheless, during the same process potentially harmful reactive oxidation products are also formed. The first oxidation product of quercetin is a semiquinone radical (Metodiewa et al., 1999). This radical is unstable and rapidly undergoes a second oxidation reaction that produces another quinone (quercetin-quinone, QQ) (Metodiewa et al., 1999). Since QQ can react with proteins, lipids and DNA, it is responsible for protein and DNA damage as well as lipid peroxidation.

The oxidative decomposition of quercetin by hydroperoxidase activity of lipoxygenase has been reported, suggesting that in the presence of the lipoxygenase/H_2O_2 system quercetin is oxidized to a quinoid product. It is remarkable that this behavior is not shown by naringenin or resveratrol, other bioactive antioxidant phenolics, probably due to the different redox potentials of these compounds (Pinto & Macias, 2005).

UV-Vis spectroscopy is a widespread and commonly used technique that has been successfully used for the determination of catalytic mechanism, including enzyme-substrate/inhibitor interaction profiles. In this report, we used UV-Vis spectroscopy to help elucidate possible interacting/inhibitory effect of quercetin with soybean LOX-1, oxidizing or not the linoleic acid. The UV-Vis spectral analysis of the mixture quercetin and LOX-1 and the kinetic parameters (K_m and V_{max}) of LOX-1-catalyzed oxidation of linoleic acid in the absence and presence of different quercetin concentrations revealed that:

i.   quercetin (λ_{max}= 370 nm) was oxidized to a new compound having λ_{max}= 321 nm by LOX-1 in absence of substrate.
ii.   a mixed inhibition occurred for quercetin concentrations in the range of 10~50 μM.
iii. at 100 μM, the highest quercetin concentration tested, the K_{mapp} decreased by half while reaction rate increased indicative of a cooxidation of quercetin in addition to the LOX classical reaction illustrating the switch in quercetin's role from inhibitor towards substrate.
iv.   quercetin concentration significantly affected its partitioning level as a substrate or an inhibitor of LOX.
v.   the ratio substrate:inhibitor might be a factor determining the type of inhibition observed in the case of lipoxygenase and quercetin interaction.

2. Biochemical analysis of LOX interaction with quercetin

2.1 UV-Vis spectroscopy

When a sample of an unknown compound is exposed to light, certain functional groups within the molecule absorb light of different wavelengths. In UV/Visible Spectroscopy, the term chromophore is used to indicate a functional group that absorbs electromagnetic radiation, usually in the UV or visible region. The type of functional groups that absorb ultraviolet light can be conjugated species, such as alkenes, aromatics, etc., making UV/Visible spectroscopy useful for distinguishing conjugated dienes from conjugated trienes, and so forth (Perkampus, 1992).

The reference beam in the spectrometer travels from the light source to the detector without interacting with the sample. The sample beam interacts with the sample exposing it to ultraviolet light of continuously changing wavelength. When the emitted wavelength corresponds to the energy level which promotes an electron to a higher molecular orbital, energy is absorbed. The detector records the ratio between reference and sample beam intensities (I_0/I). At the wavelength where the sample absorbs a large amount of light, the detector receives a very weak sample beam. Once intensity data has been collected by the spectrometer, it is sent to the computer as a ratio of reference beam and sample beam intensities. The computer determines at what wavelength the sample absorbed a large amount of ultraviolet light by scanning for the largest gap between the two beams (Perkampus, 1992).

When a large gap between intensities is found, where the sample beam intensity is significantly weaker than the reference beam, the computer plots this wavelength as having the highest ultraviolet light absorbance when it prepares the ultraviolet absorbance spectrum. Once the spectrometer has collected data from sample exposure to the UV beam, the data is transmitted to an attached computer which processes the intensity/wavelength data to produce an absorbance spectrum (Perkampus, 1992).

Various techniques have been devised for the determination of lipoxidase activity, including colorimetric, polarographic and spectrophotometric methods (Holman, 1955). Firstly, the spectrophotometric method was developed after Holman and Burr (1945) and Bergström (1946). They had independently observed an increase in ultraviolet light absorption, at 234 nm, when lipoxidase acted upon essential fatty acids. The increase in UV-peak absorption was then related to the amount of peroxide formation which, was found to be proportional to time and to enzyme concentration (Tappel, 1962; Theorell et al., 1944). The polyunsaturated fatty acid is solubilized by the addition of a detergent, and with this soluble substrate the activities of purified and crude lipoxidase are demonstrated over a wide range of pH (Tappel, 1962).

In the '50s the research done for the elucidation of lipoxygenase activity was strongly linked to the UV-Vis spectrometry and until today the UV-Vis spectrometry is an essential tool in probing the lipoxygenase activity. Thus in 1952 in an article published in the Journal of Biological Chemistry, Tappel et al. (1952), using a Beckman DU quartz spectrophotometer that was equipped with a temperature-controlled cell compartment, showed that with sodium linoleate as a substrate, under suitable conditions the reaction velocity was linear with respect to the enzyme concentration and the reaction did not show an induction period. With methyl linoleate the reaction velocity was not directly proportional to enzyme concentration unless a Tween preparation was added.

Antioxidants inhibited linoleate oxidation as a result of a direct effect on lipoxidase and of a preferential oxidation of the antioxidant. Rapid oxidation of nordihydroguaiaretic acid under suitable conditions was obtained in the absence of net linoleate oxidation, the linoleate having a function analogous to that of a coenzyme (Tappel et al., 1952).

The inactivation of soybean lipoxygenase during oxygenation of fatty acid substrates was first described by Theorell et al. (1944). It was shown that velocity of the lipoxygenase-catalyzed reaction decreases as a linear function of substrate utilization with all substrates tested.

2.2 UV-Vis spectra of LOX reaction with quercetin

We demonstrated by UV-Vis spectroscopy that pH values may influence the molecular interactions between soybean LOX-1 and quercetin, and especially the alkaline pH favors the ionic display of quercetin in order to interact with LOX better (Chedea et al., 2006). After 60 min of incubation of soybean LOX and quercetin (50 μM) the UV-Vis spectra showed the formation of a new product (λ_{max} = 321 nm) indicating the formation of an intermolecular complex between LOX and quercetin (Fig. 6) (Chedea et al., 2006).

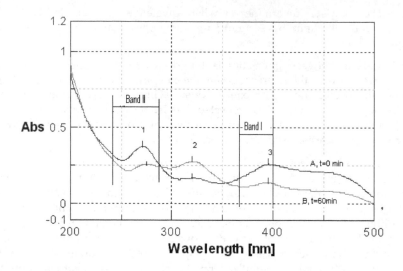

Fig. 6. The absorption spectra of the mixture soybean LOX-1 and quercetin (50 µM) at the initial moment (min 0), blue line, and after 60 min of incubation, green line. For measurements 160 µl standard LOX-1 (2300 enzymatic units/ml) and quercetin to the final concentration of 50 µM were added to 790 µl 0.2 M phosphate buffer pH 9. The spectra were registered on a Jasco-V 500 spectrophotometer and the blank contained 950 µl buffer and 50 µl ethanol (Chedea et al., 2006).

Fig. 6 shows the spectra of the mixture of LOX and quercetin. Three peaks (1, 2, 3) registered at t=0 and t=60 min of incubation were identified. Peak 1 corresponds to Band II, peak 3 to Band I, both bands characteristic for flavonoids, while peak 2 indicated the formation of a new compound as the result of reaction between lipoxygenase and quercetin.

In the spectral range of 240-450 nm, flavone and their hydroxy substituted derivatives show two main absorption bands commonly referred to as Band I (300-400 nm) and Band II (240-280 nm) (Mabry et al., 1970). Band I is supposed to be associated with the light absorption of the cinnamoyl system (B+C ring), and Band II with the absorption of the benzoyl moiety formed by the A+C ring (Fig. 7) (Zsila et al., 2003).

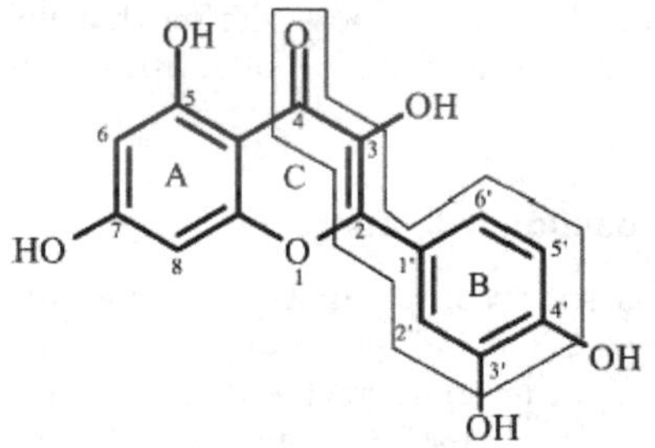

Fig. 7. Chemical structure of quercetin. Frame highlights the cinnamoyl part of the molecule (Zsila et al., 2003).

The results of Chedea et al. (2006) are confirmed by the study of Ha et al. (2010). They observed the decrease in the absorbance of the band centered at 272 and 386 nm with the concomitant increase in the absorbance at 330 nm, and the presence of two isosbestic points at 286 and 357 nm, respectively, which is also in agreement with the report of Takahama (Takahama, 1985), and the hypsochromic shift likely indicates that quercetin was oxidized (Ha et al., 2010). The absorbance at 385 nm mainly results from the $n \rightarrow \pi^*$ transition, and hence, the shift may be caused by structural changes including the ketone moiety, besides the change of the catechol on the ring B to the corresponding o-quinone (Ha et al., 2010). In the study of Pinto and Macias, when quercetin is incubated in the presence of lipoxygenase and linoleic acid, a decrease of the band centered at 375 nm is produced, together with a slight increase of absorbance in the region of 330-340 nm (Pinto & Macias, 2005).

Lipoxygenase inhibition, as a result of intermolecular interaction between the enzyme and quercetin, is dependent on the quercetin's concentration. As the concentration of inhibitor increases the formation of a new compound as a result of this reaction increases as well, showing that a greater concentration of quercetin determines a more intense interaction between lipoxygenase and inhibitor (Vicaş et al., 2006). Quercetin exhibited a dose-dependent inhibitory effect, and the lipoxygenase-catalyzed oxidation of linoleic acid to 13-HPOD was inhibited with an IC_{50} value of 4.8 ± 4 µM (Ha et al., 2010).

Pinto at al. (2011) have shown also the existence of the interaction between lipoxygenase and quercetin. They also investigated the formation of an intermolecular complex between quercetin and lipoxygenase (Pinto et al., 2011). The acting forces between lipoxygenase and quercetin include mainly hydrogen bond and van der Waals, electrostatic and hydrophobic forces (Pinto et al., 2011).

2.3 Enzyme kinetics

Enzyme kinetics is the investigation of how enzymes bind substrates and turn them into products. The rate data used in kinetic analyses are obtained from enzyme assays. In 1913 Leonor Michaelis and Maud Menten proposed a quantitative theory of enzyme kinetics, which is referred to as Michaelis-Menten kinetics. Their work was further developed by G. E. Briggs and J. B. S. Haldane, who derived kinetic equations that are still widely used today.

The major contribution of Michaelis and Menten was to think of enzyme reactions in two stages. In the first, the substrate binds reversibly to the enzyme, forming the enzyme-substrate complex. This is sometimes called the Michaelis-Menten complex in their honor. The enzyme then catalyzes the chemical step in the reaction and releases the product (Rogers & Gibon, 2009).

To find the maximum speed of an enzymatic reaction, the substrate concentration is increased until a constant rate of product formation is seen. Saturation happens because, as substrate concentration increases, more and more of the free enzyme is converted into the substrate-bound ES form. At the maximum velocity (V_{max}) of the enzyme, all enzyme active sites are saturated with substrate, and the amount of ES complex is the same as the total amount of enzyme.

However, V_{max} is only one kinetic constant of enzymes. The amount of substrate needed to achieve a given rate of reaction is also important. This is given by the Michaelis-Menten constant (K_m), which is the substrate concentration required for an enzyme to reach one-half its maximum velocity. For finding the substrate concentration required for an enzyme to reach one-half its maximum velocity, the reaction speed is measured at different substrate concentrations. Each enzyme has a characteristic K_m for a given substrate, and this can show how tight the binding of the substrate is to the enzyme (Rogers & Gibon, 2009).

The effect of substrate concentration ([S]) on activity is usually expressed using a Michaelis-Menten plot, such as the one shown below, and enzymes which generate such a plot are said to obey Michaelis-Menten kinetics. Michaelis-Menten plots show three distinct regions which correspond to reaction order. At low [S], the reaction accelerates as more substrate is added, reflecting first-order kinetics. At high [S], the concentration of enzyme becomes limiting, and additional substrate cannot accelerate the reaction. This situation is known as zero-order kinetics. Finally, there is a transition period between first order and zero order where kinetics are mixed (http://wiz2.pharm.wayne.edu/biochem/enz.html).

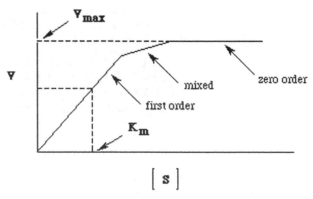

Fig. 8. A Michaelis-Menten plot.

If one draws a line across from the level (zero order) region of the plot to the Y-axis, this data point is V_{max}, the maximum rate of reaction for a given concentration of enzyme. The second kinetic constant is also derived by drawing a line from the Y-axis at $1/2$ V_{max} to the curve, and then down to the X-axis. Each substrate will generate a unique K_m and V_{max} for a given enzymatic process.

A standard equation used to express the kinetic constants under the Michaelis-Menten hypothesis is aptly called the Michaelis-Menten equation, and is shown below. Later, two other investigators rearranged this equation to generate a second useful equation, the Lineweaver-Burk equation, also shown below.

Two things should be noticed about the Lineweaver-Burk equation: first, it is in the form y = mx + b, and as such, a plot of this equation will generate a straight line for enzymes obeying simple Michaelis Menten kinetics. In addition, the x and y values for the plot are both inverted, and as such, the plot is often referred to as the double reciprocal plot. The Lineweaver-Burk plot has two advantages over the Michaelis-Menten plot, in that it gives a more accurate estimate of V_{max}, and it gives more accurate information about inhibition as

The Michaelis-Menton Equation

$$V = \frac{Vmax\,[S]}{Km + [S]}$$

The Lineweaver-Burke Equation

$$\frac{1}{V} = \frac{K_m}{V_{max}}\frac{1}{[S]} + \frac{1}{V_{max}}$$

well. A typical Lineweaver-Burk plot appears below. Note that V_{max} is derived from the y-intercept, and K_m can be derived either from the slope, or from extrapolating the line to the negative X-axis (http://wiz2.pharm.wayne.edu/biochem/enz.html).

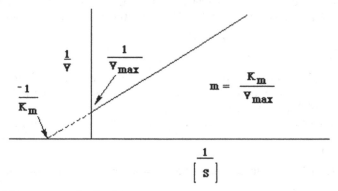

Fig. 9. A typical Lineweaver-Burk plot.

2.3.1 Enzyme inhibition

There are two main classes of enzyme inhibitors, reversible and irreversible, that are differentiated by the magnitude of their affinity for enzyme. Reversible enzyme inhibitors bind and dissociate with their enzyme in an equilibrium process. Irreversible inhibitors bind tightly to an enzyme to form an essentially permanent complex. Reversible inhibitors can be classified as competitive, mixed, or noncompetitive inhibitors. If the detailed mechanism of inhibition is known, then the classification can be made by identifying where on the enzyme the inhibitor binds, or the order with which it binds, relative to substrate. Alternatively, a determination of simple kinetic parameters can generally be used to classify the inhibitor (http://www.wiley.com/college/pratt/0471393878/student/animations/enzyme_inhibitio n/index.html)

2.3.1.1 Competitive inhibitors

Competitive inhibitors compete with substrate for an enzyme's active site, lowering the enzyme's likelihood of binding substrate and slowing the observed reaction velocity. Kinetic studies can be used to determine the type and potency of inhibition for an unknown inhibitor. Typical steady-state kinetic experiments can be performed where reaction velocity is measured in the presence of varying concentrations of substrate. If inhibitor is then added, and the data shows an increase in K_m, yet the V_{max} is unaffected, this is the signature of a competitive inhibitor (http://www.wiley.com/college/pratt/0471393878/student/animations/enzyme_inhibition/index.html)

2.3.1.2 Mixed inhibition

A mixed inhibitor binds to a site on the enzyme and interferes with both apparent substrate affinity and catalytic turnover, thus affecting the observed K_m for the enzyme-catalyzed reaction. Mixed inhibitors do not bind directly in the active site, and therefore do not block substrate binding, but instead bind at sites that can be proximal or distal from the active site. Mixed inhibitors can therefore bind to free enzyme prior to substrate, distorting the active site to a nonoptimal conformation for catalysis. The inhibitor-distorted active site has trouble converting the substrate to product before it dissociates, resulting in a lowered apparent substrate binding affinity. Steady-state experiments performed in the presence of a mixed inhibitor demonstrate an increase or decrease in K_M, and a decrease in V_{max} (http://www.wiley.com/college/pratt/0471393878/student/animations/enzyme_inhibitio n/index.html)

2.3.1.3 Noncompetitive inhibition

Noncompetitive inhibition is a special case of mixed inhibition where the affinity of inhibitor for E and ES is the same. Steady-state experiments performed in the presence of a noncompetitive inhibitor demonstrate a decrease in V_{max}, yet K_m is unaffected (http://www.wiley.com/college/pratt/0471393878/student/animations/enzyme_inhibitio n/index.html)

2.4 The influence of pure quercetin on sodium linoleate oxidation by pure soybean LOX-1

2.4.1 The influence of quercetin on the LOX-1 oxidation of sodium linoleate for different experimental protocols

It is of great interest to check if the oxidative decomposition of quercetin by lipoxygenase is produced in the presence of linoleic acid. It is known that the hydroperoxidase activity of lipoxygenase produces the cooxidation of suitable electron donors in the presence of the hydroperoxides of linoleic or arachidonic acid, the natural substrates for this enzyme (Pinto & Macias, 2005). When linoleic acid is used as the substrate the primary LOX dioxygenation product obtained is (9Z,11Z,13S)-hydroperoxyoctadeca-9,11-dienoic acid (13-HPOD) (Grechkin, 1998).

Dioxygenase activity of lipoxygenase was measured by recording the increase in absorbance at 234 nm (formation of 13-HPOD), the incubation mixture containing 50 µl quercetin in ethanol to a final concentration of 100 µM, 8.4 µl sodium linoleate at different concentrations (2.2 mM, 4.39 mM, 6.6 mM and 20 mM) and 160 µl lipoxygenase (2300 units/ml) in 782 µl 0.2 M borate buffer pH 9. The final substrate concentrations were 18.5 µM, 36.9 µM, 55.4 µM respectively 168 µM. The reaction components were mixed following three protocols as Fig. 10 indicates. The differences between these protocols are given by the order in which the enzyme, the substrate and the inhibitor were mixed and the incubation time.

As already presented in section 2.2, lipoxygenase in the absence of linoleic acid interacts with quercetin, thus Protocol 1 was designed to establish how the inhibition of substrate oxidation occurs when LOX-1 is initially incubated with the inhibitor. On the hand, Protocol 2 follows the classical way of LOX assay where the enzyme reacts with the substrate in the presence of quercetin. In the typical LOX reaction, the oxidation of the iron atom occurs with consumption of one molecule of fatty acid hydroperoxide (Zheng & Brash, 2010), therefore, the objective of the protocol was to explore the possibility of quercetin reacting with the LOX reaction product, the 13-hydroperoxide. Finally, the objective of Protocol 3 was to check for the existence of competition between the substrate and inhibitor in a reaction environment where both linoleic acid and quercetin-also oxidized by lipoxygenase-exist.

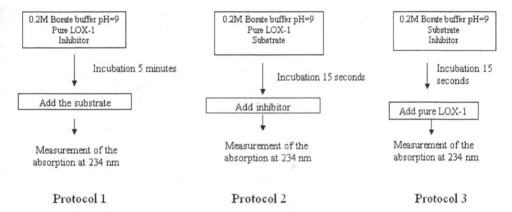

Fig. 10. Schematic representation of experimental protocols (1,2,3).

Fig. 11. presents comparatively the inhibitory effect of quercetin 100 µM on the oxidation of sodium linoleate by LOX-1 using all 3 experimental protocols. The kinetic curves of the reaction in the presence of quercetin are compared with the one of lipoxygenase without inhibitor (red curve).

It can be seen from Fig. 11 that the strongest inhibition is reached in the case of protocol 2 (pink line) when the inhibitor was added after the incubation for 15 seconds of sodium linoleate 55.4 µM with the enzyme. For protocol 1 the typical Michaelis-Menten kinetic plot shape was maintained like for the LOX reaction with substrate without inhibitor (Fig. 11 blue curve vs. red plot). A different shape has the plot in the case of protocols 2 (pink line) and 3 (green line). At low substrate concentrations for both protocols, 2 and 3 (Fig. 11 curves green and pink), the kinetic plots start with a burst phase-a very fast increase of the reaction velocity. It follows a decrease which is fast in the case of protocol 3 (green curve) and slower in the case of protocol 2 (pink curve). For protocol 2 the kinetic curve continues with a slight increase (starting with 55.4 µM concentration of substrate), which in our situation didn't reach the plateau phase. In the case of protocol 3 a new increase of the reaction rate -not so high compared with the first burst- follows at substrate concentration between 36.9 µM and 55.4 µM, continuing with a slight decrease at concentrations of substrate higher than 55.4 µM.

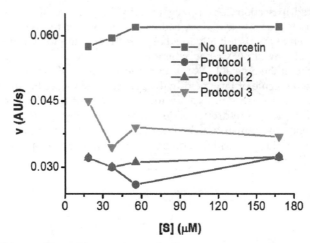

Fig. 11. The inhibitory effect of quercetin on the lipoxygenase reaction according to protocol 3- dark green line; to protocol 2- pink line; to protocol 1- blue line. The red curve represents the kinetic without quercetin. The reaction components, 782 µl 0.2 M borate buffer, (pH = 9.0), 160 µl of LOX solution (2300 U enzyme/ml) 8.4 µl of sodium linoleate, of different concentrations (2.2 mM, 4.39 mM, 6.6 mM and 20 mM) having the final substrate concentration in the reaction mixture of 18.5 µM, 36.9 µM, 55.4 µM and 168 µM) and 50 µl quercetin in ethanol to the final concentration of 100 µM were added and mixed as Fig. 10. presents. The absorption at 234 nm (A_{234}) against the blank was recorded.

When both the substrate and the inhibitor were mixed and left for few seconds and then LOX was added the kinetic curve (Fig. 11 protocol 3), the first phase of fast increase shows that the hydroperoxides are formed as oxidation product of the linoleic acid even though the quercetin is present in the reaction mixture for very low substrate concentrations. LOX is partially inhibited by the quercetin at 36.9 µM substrate concentration but then, at higher concentrations of substrate the enzyme becomes active again, but not as much as in the case of very low substrate concentration. Rapid inhibition followed by time dependent inactivation of soybean LOX-1 was also observed by Sadik et al. when quercetin was added to the reaction set-up after the substrate (Sadik et al., 2003).

When LOX was incubated with the substrate for 15 seconds and then the quercetin was added, the kinetic plot shows that LOX oxidizes the substrate having the highest product yield for very low substrate concentrations, that quercetin partially inhibits the reaction at linoleate concentrations between 30 µM and 55 µM.

Careful examination of the three protocols leads to speculation the quercetin-based inhibition of the soybean LOX-1 oxidation of linoleic acid does not follow the typical competitive inhibition model under the experimental conditions used (comparatively, protocol 3 gave the lowest inhibition than the other protocols tested). In Protocol I, it was observed that the new compound, a LOX-1-catalyzed by-product of quercetin, effectively inhibited the linoleic acid's oxidation. In the case where quercetin is added the last, the LOX reaction inhibition pattern suggested that at certain substrate concentrations (between 45 µM and 75 µM) quercetin would react with the hydroperoxide responsible for the initiation of the LOX cycle, and thus implying the possibility of the quercetin undergoing non-enzymatic oxidation leading to the reaction inhibition.

2.4.2 Determination of kinetic parameters K_m and V_{max} of soybean LOX-1 standard towards linoleic acid as substrate

The kinetic parameters for LOX-1 oxidation of sodium linoleate were calculated as control values and compared with LOX-1 oxidation of sodium linoleate in the presence of different concentrations of quercetin. For the measurements of pure LOX activity, to 832 µl 0.2 M borate buffer, (pH = 9.0), 160 µl of LOX solution (2300 U enzyme/ml) and 8.4 µl of sodium linoleate of different concentrations (2.2 mM, 4.39 mM, 6.6 mM and 20 mM), having the final substrate concentration in the reaction mixture of 18.5 µM, 36.9 µM, 55.4 µM and 168 µM, were added. The absorption at 234 nm (A_{234}) against the blank was recorded. The blank contained a mixture of 840 µl 0.2 M borate buffer, (pH = 9.0) and 160 µl of LOX solution (2300 U enzyme/ml). In order to obtain the Lineweaver-Burk plot, the v, 1/[S] and 1/v were calculated for different substrate concentrations (from 18.5- 168 µM).

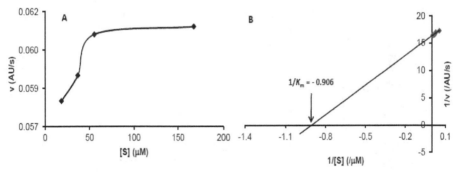

Fig. 12. Michaelis-Menten plot (A) and Lineweaver-Burk plot (B) for different sodium linoleate concentrations oxidized by standard LOX-1. For the measurements of pure LOX activity, to 832 µl 0.2 M borate buffer, (pH = 9.0), 160 µl of LOX solution (2300 U enzyme/ml) and 8.4 µl of sodium linoleate of different concentrations (2.2 mM, 4.39 mM, 6.6 mM and 20 mM), having the final substrate concentration in the reaction mixture of 18.5 µM, 36.9 µM, 55.4 µM and 168 µM, were added. The absorption at 234 nm (A_{234}) against the blank was recorded. The blank contained a mixture of 840 µl 0.2 M borate buffer, (pH = 9.0) and 160 µl of LOX solution (2300 U enzyme/ml) (Chedea et al., 2008). Excel program was used to draw the graphs.

The kinetic parameters calculated for LOX-1 oxidizing sodium linoleate show that the enzyme has a great affinity towards substrate (K_m = 1.1 µM) and that its oxidation is fast as well (V_{max}= 2.7 µMs^{-1}).

2.4.3 Kinetic parameters for LOX-1 oxidizing sodium linoleate in presence of different quercetin concentrations

The reaction velocities in the case of this first protocol for the three quercetin concentrations were calculated as the ratio of absorption and correspondent time registered at 234 nm. For the absorption measurements of pure LOX activity to 782 μl 0.2 M borate buffer, (pH = 9.0), 160 μl of LOX solution (2300 U enzyme/ml) 50 μl quercetin (10 μM, 50 μM and 100 μM final concentrations) and 8.4 μl of sodium linoleate of different concentrations (2.2 mM, 4.39 mM, 6.6 mM and 20 mM), having the final substrate concentration in the reaction mixture of 18.5 μM, 36.9 μM, 55.4 μM and 168 μM, were added. The absorption at 234 nm (A_{234}) against the blank was recorded. The blank contained a mixture of 840 μl 0.2 M borate buffer, (pH = 9.0) and 160 μl of LOX solution (2300 U enzyme/ml).

The inhibitory effect of quercetin at different concentrations (100, 50 and 10 μM) is presented in Fig. 13.

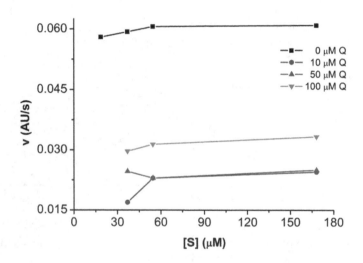

Fig. 13. Lipoxygenase inhibition by quercetin at different concentrations (10 μM, 50 μM and 100 μM). Protocol 1 (Fig. 10) was followed for measurements as indicated in the legend of Fig. 11.

Plotting the reaction velocity in function of substrate concentration it can be seen that the classical Michaelis-Menten shape of the curve is not registered in the case of LOX reaction inhibition by quercetin at 50 μM. For this reason, it was taken into calculation a larger range of substrate and quercetin concentration and for those the reaction velocity was calculated.

The kinetic curves representing the reaction velocity of the sodium linoleate oxidation by LOX-1, in presence of different concentrations of quercetin (Fig. 13) show that quercetin has an inhibitory effect on lipoxygenase for all the concentrations tested. To obtain these results, the reaction components were mixed as protocol 1 indicates, the enzyme being incubated with quercetin for 5 minutes and then the substrate is added. From Fig. 14 it can be seen that none of the kinetic plots follows the classical Michaelis-Menten shape. To get more information about the quercetin inhibitory action on soybean LOX-1, the kinetic parameters were calculated.

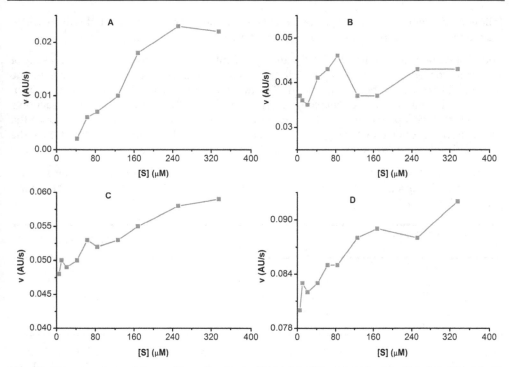

Fig. 14. Kinetic plots of the sodium linoleate (336 µM, 252 µM, 168 µM, 126 µM, 84 µM, 63 µM, 42 µM, 21 µM, 10.5 µM and 5.25 µM), oxidation by LOX-1, in presence of different concentrations of quercetin 10 µM (A), 25 µM (B), 50 µM (C) and 100 µM (D).

Fig. 14 indicates different behaviours of LOX-1 quercetin inhibition determined by the substrate's and inhibitor's concentrations. For quercetin at 100 µM (Fig. 14D), 50 µM (Fig. 14C), at low concentrations of substrate (up to 21 µM a fast increase of the reaction velocity is recorded, followed by a slight decrease and increase again (concentration of substrate between 21 µM and 42 µM). This "oscillatory" behaviour continues at higher substrate concentrations in the case of quercetin at 100 µM and 50 µM with a tendency of increasing the reaction's velocity at concentrations of substrate higher than 336 µM for 100 µM quercetin and of reaching the plateau phase for 50 µM quercetin.

A fast decrease followed by an increase until the maximum velocity, characterises the kinetic plot of LOX-1 oxidizing the sodium linoleate at low substrate concentrations up to 84 µM (Fig. 14B) in the presence of quercetin 25 µM. At this concentration of substrate (84 µM) the reaction velocity reaches its maximum. It follows a significant decrease of the reaction velocity at substrate concentrations between 84 µM and 134.4 µM, the reaction rate increasing again and having the tendency to follow the classical Michael-Menten kinetic shape till at 252 µM, substrate's concentration. At this point a slight decrease of the reaction rate is registered, showing that for quercetin 25 µM the inhibition is given by the high substrate concentration as well, when the sodium linoleate has a concentration higher than 252 µM (Fig. 14B). The same behaviour is registered in the case of quercetin 10 µM at linoleate concentrations higher than 252 µM (Fig. 14A). For quercetin 10 µM the kinetic plot follows the tendency of the classical Michaelis-Menten shape, with a slight decrease in reaction velocity at concentrations of substrate between 63 µM and 126 µM.

The observation that diminution of the degree of inhibition occurred by the decreasing of substrate concentration may imply that the substrate oxidation also plays a role in the process of LOX-1 inhibition by quercetin. This involvement may be one of the initial steps in the LOX inhibition, where quercetin is non-enzymatically oxidized by the reaction's product, the hydroperoxide, resulting in the formation of the quercetin o-quinone. Once the "activation" of quercetin through its oxidation is triggered, it can be assumed that the enzymatic inhibition would proceed via a quercetin oxidation product.

K_m, K_{mapp} and V_{max} calculated from the Lineweaver-Burk plots, for LOX-1 oxidizing sodium linoleate in the absence and presence of quercetin at 10 µM, 25 µM, 50 µM and 100 µM are presented in Table 1.

Quercetin conc. µM	K_{mapp} (µM)	V_{max} (µMs⁻¹/s)	Type of inhibition
100	0.55	3.8	-
50	0.83	2.4	mixed
25	0.78	1.8	mixed
10	261.78	0.5	mixed
0	1.103	2.7	

Table 1. K_m, K_{mapp} and V_{max} calculated from the Lineweaver-Burk plots, for LOX-1 oxidizing sodium linoleate in the absence and presence of quercetin at 10 µM, 25 µM, 50 µM and 100 µM.

The determined K_{mapp} and V_{max} show a mixed inhibition for quercetin concentrations in the range of 10-50 µM. For 100 µM the K_{mapp} is half decreased and the reaction rate increases, reaching its highest value.

In a previous study LOX activity was measured spectrophotometrically at 234 nm using 15 mU of enzyme in the presence of 100 µM linoleic acid in 50 mM potassium phosphate buffer, pH 7.5. For each scan 2 µl of quercetin solution was added to 3 mL lipoxygenase solution to give a final concentration in the range 0.8 to 4.0 µM and an inhibitory effect concentration dependent was detected (Pinto et al., 2011). This effect shows the characteristics of a competitive mechanism as it was deduced from Lineweaver-Burk plot (Pinto et al., 2011). On the basis of the competitive inhibition detected, the interaction should be located near or at the catalytic site (Pinto et al., 2011). The results obtained from the evaluation of three dimensional florescence spectra suggest a conformational modification of the protein in the region of the coupling with quercetin (Pinto et al., 2011).

The degree of inhibition was paradoxically diminished with decreasing substrate concentration which reveals an unusual mode of the inhibitory effect (Sadik et al., 2003). In this case, a competitive type of inhibition appears to be excluded (Sadik et al., 2003). The rabbit reticulocyte 15-LOX-1 samples were pre-incubated for 2 min at 20° with 10 µM quercetin and the reactions were started by addition of 0.265 mM potassium linoleate and the formation of conjugated dienoic fatty acids was recorded spectrophotometrically at 234 nm (Sadik et al., 2003). Ha et al. report that quercetin inhibited soybean LOX-1 in a non-classical manner. The progress curves of O_2 consumption showed that quercetin inhibited soybean LOX-1 by a slow-binding inhibition mechanism (Ha et al., 2010).

3. Conclusions

The aim of this study was to show how a widespread analysis tool like UV-Vis spectroscopy shapes the enzyme inhibition research having as an example the lipoxygenase interaction with quercetin. Almost all therapeutic drugs are enzyme inhibitors, from old medicine box standards such as aspirin and penicillin to the newest compounds used to treat HIV infection. Understandably, enzyme kinetics plays an outstanding role in this effort to produce effective therapeutics, for kinetic studies can quantify the degree that inhibitors inactivate or slow down the targeted enzyme's catalytic rate and describe its potential efficacy as a drug.

Since its characterisation in 1947 lipoxygenase is strongly related to the UV-Vis spectrometry as a valuable tool for its activity assay. Due to its implications in food chemistry and medicine LOX inhibition attracted up to date, the interest of researchers in the field. A highly functionalized flavonoid, quercetin proves to be a potent inhibitor of lipoxygenase acting both as a substrate and a source of inhibition, quercetin seems to play an antinomic role (Fiorucci et al., 2008). The partitioning level between quercetin as a substrate or as an inhibitor is dependent on its concentration. Different reversible inhibition schemes (competitive, noncompetitive, or mixed) as well as inhibition via reduction of the enzyme-bound radical intermediate have been considered to explain the activity of quercetin as LOX inhibitor. Moreover, heterogeneity in the interpretation of the experimental results of the inhibition processes, for example concerning kinetic data, prevents converging toward a general way of inhibition (Fiorucci et al., 2008). The ratio substrate:inhibitor might be a factor determining the type of inhibition observed in the case of lipoxygenase and quercetin interaction.

In our present study the UV-Vis spectra show the oxidation of quercetin and the formation of a new compound. The absorption maximum of this new formed molecule is centered around λ_{max}= 321 nm, different from quercetin (λ_{max}= 370 nm), suggesting loss of π-electron delocalisation, i.e. interruption of the quercetin B and C ring π-bond extended conjugated system (Bors et al., 1990; Abou Samra et al., 2011). The determined K_{Mapp} and v_{max} show a mixed inhibition for quercetin concentrations in the range of 10-50 μM. For the highest quercetin concentration tested, 100 μM, the K_{mapp} decreased by half but the reaction rate increases which might indicate a cooxidation of quercetin besides the LOX classical reaction, proving the switch in quercetin role from inhibitor towards substrate. Recent literature data show that quercetin itself inhibits the LOX reaction and also its oxidation products are inhibitors of LOX oxidation. For instance, quercetin may first act as a lipoxygenase inhibitor by reducing the ferric form of the enzyme to an inactive ferrous form, and then, the oxidized metabolite becomes a more potent inhibitor (Ha et al., 2010).

In our study, the Michaelis-Menten (M-M), Lineweaver-Burk kinetics and UV spectral analysis have detected both the mixed inhibition and cooxidation of quercetin during soybean LOX-1-mediated metabolism of polyunsaturated fatty acid, linoleic acid. Though not yet convincingly proven, the mechanisms underlying the distinct kinetic behaviors are, at least empirically, believed to be due to the existence of divergent interactions between the quercetin molecules and/or its intermediates and the active site of the enzymes. Furthermore, the atypical kinetics might be interpreted using the model of two or possibly more binding regions within the enzyme active site(s) for the quercetin and intermediates. The finding that LOX can turn different compounds, like quercetin and epigallocatechin gallate, into simple catechol derivatives (with one aromatic ring only) might be of importance as an additional small piece of a "jigsaw puzzle" in the much bigger picture of drug metabolism (Borbulevych et al., 2004). However, the interactions of these flavanols with LOX can be more complicated than simply blocking the access to the enzyme's active site as observed in this study. Therefore, it warrants future endeavours to thoroughly understand thus reliably predict interaction mechanism between the LOX proteins and therapeutic agents at the molecular level. This will be important in fully understanding the exact role of lipoxygenase inhibition by quercetin in therapy targeting and possibly identifying new bioactive molecules which would be used as drugs.

4. Acknowledgment

V.S. Chedea is a Japan Society for the Promotion of Science (JSPS) postdoctoral fellow.

5. References

Abou Samra, M., Chedea, V. S., Economou, A., Calokerinos, A. & Kefalas, P. (2011). Antioxidant/Prooxidant properties of model phenolic compounds. Part I: Studies on equimolar mixtures by chemiluminescence and cyclic voltammetry, *Food Chemistry*, Vol. 125 (2), pp. 622-629.

Aharony D. & Stein R.L. (1986). Kinetic mechanism of guinea pig neutrophil 5-lipoxygenase. *Journal of Biological Chemistry*, Vol. 261(25), pp. 11512-11519, ISSN: 1083-351X.

Ariel, A., Li, P.L., Wang, W., Tang, W.X., Fredman, G., Hong, S., Gotlinger, K.H. & Serhan, C.N. (2005). The docosatriene protectin D1 is produced by TH2 skewing and promotes human T cell apoptosis via lipid raft clustering. *The Journal of Biological Chemistry*. Vol. 280(52), pp. 43079–43086, ISSN: 1083-351X.

Awad, H.M., Boersma, M.G., Vervoort, J. & Rietjens, I.M.C.M. (2000). Peroxidase-catalyzed formation of quercetin quinine methide-glutathione adducts. *Archives of Biochemistry and Biophysics*, Vol. 378, pp. 224–233, ISSN: 0003-9861.

Awad, H.M., Boersma, M.G., Boeren, S., van Bladeren, P.J., Vervoort, J. & Rietjens, I.M.C.M. (2001). Structure-activity study on the quinone-quinone methide chemistry of flavonoids. *Chemical Research in Toxicology*, Vol. 14, pp. 398-408, ISSN: 0893-228X.

Beecher, G. R., Warden, B. A. & Merken, H. (1999). Analysis of tea polyphenols. *Proceedings of the Society for Experimental Biology and Medicine*, Vol. 220, No. 4, pp. 267–270.

Bergström, S. (1946). On the oxidation of linoleic acid with lipoxidase. *Arkiv Kemi Mineralogi Och Geologi. Band* 21 A. No. 15.

Bhattacharya, S., Mathew, G., Jayne, D.G., Pelengaris, S. & Khan, M. (2009). 15-Lipoxygenase-1 in colorectal cancer: A review. *Tumor Biology*, Vol. 30, pp. 185–199, ISSN 1010-4283.

Bischoff, S.C. (2008). Quercetin: potentials in the prevention and therapy of disease. *Current Opinion in Clinical Nutrition & Metabolic Care*, Vol. 11, pp. 733-740, ISSN: 1473-6519.

Boersma, M.G., Vervoort, J., Szymusiak, H., Lamanska, K., Tyrakowska, B., Cenas, N., Segura-Aguilar, J. & Rietjens, I.M.C.M. (2000). Regioselectivity and reversibility of the glutathione conjugation of quercetin quinone methide. *Chemical Research in Toxicology*, Vol. 13, pp. 185-191, ISSN: 0893-228X.

Borbulevych, O.Y., Jankun, J., Selman, S.H. & Skrzypczak-Jankun, E. (2004). Lipoxygenase interactions with natural flavonoid, quercetin, reveal a complex with protocatechuic acid in its X-ray structure at 2.1 Å resolution. *Proteins*, Vol. 54, pp.13–19, ISSN: 0887-3585.

Bors, W., Heller, W., Michael, C., & Saran, M. (1990). Flavonoids as antioxidants: Determination of radical-scavenging efficiencies. In L. Packer & A. N. Glazer (Eds.). Methods in enzymology (Vol. 186, pp. 343–355). San Diego: Academic Press.

Borngräber, S., Browner, M., Gillmor, S., Gerth, C., Anton, M., Fletterick, R. & Kühn, H. (1999). Shape and specificity in mammalian 15-lipoxygenase active site. The functional interplay of sequence determinants for the reaction specificity. *The Journal of Biological Chemistry*, Vol.274, pp. 37345–37350, ISSN: 1083-351X.

Chawengsub, Y., Gauthier, K. M. & Campbell, W. B. (2009). Role of arachidonic acid lipoxygenase metabolites in the regulation of vascular tone, *American Journal of Physiology- Heart Circulatory Physiology*, Vol. 297, pp. H495–H507, ISSN: 0363-6135, eISSN: 1522-1539.

Chedea, V. S., Vicaş, S.I., Oprea, C. & Socaciu, C. (2006). Lipoxygense kinetic in relation with lipoxygenase's inhibition by quercetin, *Proceedings of the XXXVI Annual Meeting of the European Society for New Methods in Agricultural Research (ESNA)*, (10 – 14 September 2006, Iasi-Romania), pp. 449-456, ISBN (10) 973-7921-81-X ; ISBN (13) 978-973-7921-81-9.

Chedea, V. S., Vicaş, S. & Socaciu, C. (2008). Kinetics of soybean lipoxygenase are related to pH, substrate availability and extraction procedures, *Journal of Food Biochemistry*, Vol. 32 (2), pp. 153-172.

Chedea V.S. & Jisaka M. (2011). Inhibition of Soybean Lipoxygenases – Structural and Activity Models for the Lipoxygenase Isoenzymes Family, Recent Trends for Enhancing the Diversity and Quality of Soybean Products, Dora Krezhova (Ed.), ISBN: 978-953-307-533-4, InTech, Available from:
http://www.intechopen.com/articles/show/title/inhibition-of-soybean-lipoxygenases-structural-and-activity-models-for-the-lipoxygenase-isoenzymes-f

Choi, J., Chon, J. K., Kim, S. & Shin, W. (2008). Conformational flexibility in mammalian 15S-lipoxygenase: Reinterpretation of the crystallographic data. *Proteins*, Vol. 70, pp.1023–1032, ISSN: 0887-3585.

Coffa, G., Schneider, C. & Brash, A.R. (2005). A comprehensive model of positional and stereo control in lipoxygenases. *Biochemical and Biophysical Research Communications*, Vol. 338, pp. 87–92, ISSN: 0006-291X.

Duroudier, N.P., Tulah, A.S. & Sayers, I. (2009). Leukotriene pathway genetics and pharmacogenetics in allergy. *Allergy*, Vol. 64, pp. 823–839, ISSN: 0105-4538.

Eckl, K.M., de Juanes, S., Kurtenbach, J., Nätebus, M, Lugassy, J., Oji, V., Traupe, H., Preil, M.L., Martínez, F., Smolle, J., Harel, A., Krieg, P., Sprecher, E. & Hennies, H.C. (2009). Molecular analysis of 250 patients with autosomal recessive congenital ichthyosis: evidence for mutation hotspots in ALOXE3 and allelic heterogeneity in ALOX12B. *Journal of Investigative Dermatology*, Vol. 129, pp. 1421–1428, ISSN: 0022-202X; EISSN: 1523-1747.

Epp, N., Furstenberger, G., Muller, K. de Juanes, S., Leitges M., Hausser, I., Thieme, F., Liebisch, G., Schmitz, G. & Krieg P. (2007). 12R-lipoxygenase deficiency disrupts epidermal barrier function. *The Journal of Cell Biology*, Vol.177, pp. 173–182, ISSN: 0021-9525.

Feltenmark, S., Gautam, N., Brunnstrom, A., Griffiths, W., Backman, L., Edenius, C., Lindbom, L., Bjorkholm, M. & Claesson, H.E. (2008). Eoxins are proinflammatory arachidonic acid metabolites produced via the 15-lipoxygenase-1 pathway in human eosinophils and mast cells. *Proceedings of the National Academy Sciences USA*, Vol. 105, pp. 680–685, ISSN: 0027-8424.

Fiorucci, S., Golebiowski, J., Cabrol-Bass, D. & Antonczak, S. (2008). Molecular simulations enlighten the binding mode of quercetin to lipoxygenase-3. *Proteins*, Vol. 73, pp. 290–298, ISSN: 0887-3585.

Formica J. V., (1995). Review of the biology of quercetin and related bioflavonoids. *Food and Chemical Toxicology*, Vol. 33, No. 12, pp. 1061–1080.

Furstenberger, G.; Epp, N.; Eckl, K.M.; Hennies, H.C.; Jorgensen, C.; Hallenborg, P.; Kristiansen, K. & Krieg, P. (2007). Role of epidermis-type lipoxygenases for skin barrier function and adipocyte differentiation. *Prostaglandins Other Lipid Mediators* Vol.82 (1-4), pp.128–134, ISSN: 1098-8823.

Gardner, H.W. (1991). Recent investigations into the lipoxygenase pathway in plants. *Biochimica Biophysica Acta - Lipids and Lipid Metabolism*, Vol. 1084, No. 3, pp. 221-239.

Gillmor, S.A., Villasenor, A., Fletterick, R., Sigal, E. & Browner, M.F. (1997). The structure of mammalian 15-lipoxygenase reveals similarity to the lipases and the determinants of substrate specificity. *Nature Structural Biology* . Vol. 4, pp. 1003–1009.

Grechkin, A. (1998). Recent developments in biochemistry of the plant lipoxygenase pathway. *Progress in Lipid Research*. Vol. 37, pp. 317 –352.

Ha, T. J. & Kubo, I. (2007). Slow-binding inhibition of soybean lipoxygenase-1 by dodecyl gallate. *Journal of Agricultural Food Chemistry*. Vol. 55, pp. 446, ISSN: 0021-8561.

Ha, T. J., Shimizu, K., Ogura, T. & Kubo, I. (2010). Inhibition mode of soybean lipoxygenase-1 by quercetin. *Chemistry & Biodiversity*, Vol. 7, pp. 1893-1903, ISSN: 1612-1872.

Hamberg, M. & Samuelsson, B. (1974). Prostaglandin endoperoxides. Novel transformations of arachidonic acid in human platelets. *Proceedings of the National Academy of Science USA*, Vol. 71, pp. 3400–3404, ISSN-0027-8424.

Hersberger, M. (2010). Potential role of the lipoxygenase derived lipid mediators in atherosclerosis: leukotrienes, lipoxins and resolvins. *Clinical Chemistry and Laboratory Medicine*, Vol. 48(8), pp.1063–1073, ISSN: 1434-6621.

Heijnen C.G.M., Haenen, G.R.M.M., Van Acker, F.A.A., Van Der Vijgh W.J.F. & Bast, A. (2001). Flavonoids as peroxynitrite scavengers: the role of the hydroxyl groups, *Toxicology in Vitro*, Vol. 15, No. 1, pp. 3–6.

Holman, R. T. & Burr, G. O. (1945). Spectrophotometric studies of the oxidation of fats. IV. Ultraviolet absorption spectra of lipoxidase-oxidized fats. *Archives of Biochemistry*, Vol. 7, pp. 47-54, ISSN: 0003-9861.

Holman, R.T. (1955). Measurement of Lipoxidase Activity. In: Methods of Biochemical Analysis, D. Glick. (Ed.) Vol II, pp. 113-119, ISBN: 9780471304593, Interscience Publishers, New York.

Hollman P.C.H. & Katan, M.B. (1999). Dietary flavonoids: intake, health effects and bioavailability. *Food and Chemical Toxicology*, Vol. 37, No. 9-10, pp. 937–942.

Hover, C. G. & Kulkarni, A. P.(2000). Hydroperoxide specificity of plant and human tissue lipoxygenase: an in vitro evaluation using N-demethylation of phenothiazines. *Biochim. Biophys. Acta*, Vol. 1475, pp. 256-264.

Ivanov, I., Heydeck, D., Hofheinz, K., Roffeis, J., O'Donnell, V. B. & Kühn, H. (2010). Molecular enzymology of lipoxygenases. *Archives of Biochemistry and Biophysics.* Vol. 503, pp. 161-174.

Jakschik, B.A. & Lee, L.H. (1980). Enzymatic assembly of slow reacting substance. *Nature,* Vol. 287, pp. 51–52, ISSN: 0028-0836.

Jungbluth, G., Rühling, I. & Ternes, W. (2000). Oxidation of flavonols with Cu(II), Fe(II) and Fe(III) in aqueous media. *Journal of the Chemical Society, Perkin Transactions 2.* Vol. 2, pp. 1946-1952, ISSN: 1364-5471

Krishnamachari, V., Levine, L. H., & Pare´, P.W. (2002). Flavonoid oxidation by the radical generator AIBN: a unified mechanism for quercetin radical scavenging. *Journal of Agricultural Food Chemistry,* Vol. 50, pp. 4357.

Kubo, I., Nihei, K. & Shimizu, K. (2004). Oxidation products of quercetin catalyzed by mushroom tyrosinase. *Bioorganic & Medicinal Chemistry,* Vol. 12, pp. 5343, ISSN: 0968-0896

Kühn, H., Saam, J., Eibach, S., Holzhütter, H.G., Ivanov, I. & Walther, M. (2005). Structural biology of mammalian lipoxygenases: enzymatic consequences of targeted alterations of the protein structure. *Biochemical and Biophysical Research Communications,* Vol. 338, pp. 93–101, ISSN: 0006-291X.

Kühn, H. (2000). Structural basis for the positional specificity of lipoxygenases. *Prostaglandins & Other Lipid Mediators,* Vol. 62, pp. 255–270.

Kulkarni, A.P. (2001). Lipoxygenase – a versatile biocatalyst for biotransformation of endobiotics and xenobiotics. *Cellular and Molecular Life Sciences,* Vol. 58, pp. 1805-1825.

Loiseau, J., Benoît, L. V., Macherel, M.H. & Le Deunff, Y. (2001). Seed lipoxygenases: occurrence and functions. *Seed Science Research,* Vol. 11, pp. 199–211.

Lugli, E., Ferraresi, R., Roat, E., Troiano, L., Pinti, M. & Nasi, M. (2009). Quercetin inhibits lymphocyte activation and proliferation without inducing apoptosis in peripheral mononuclear cells, *Leukemia Research,* Vol. 33, pp. 140–150, 2009, ISSN: 0145-2126.

Mabry, T.J., Markham K.R. & Thomas M.B. (1970). The systematic identification of flavonoids, Springer-Verlag: New York. ISBN-13: 978-3540049647.

Maccarrone, M., van Aarie, P.G.M., Veldink, G.A. & Vliegenthart, J.F.G. (1994). In vitro oxygenation of soybean biomembranes by lipoxygenase-2. *Biochimica et Biophysica Acta - Biomembranes,* Vol. 1190, pp. 164–169, ISSN: 0005-2736.

Maderna, P. & Godson, C. (2009). Lipoxins: resolutionary road, *British Journal Pharmacology.* Vol. 158, pp. 947–959, ISSN: 1476-5381.

Metodiewa, D., Jaiswal, A. K., Cenas, N., Dickancaite, E. & Segura-Aguilar J. (1999). Quercetin may act as a cytotoxic prooxidant after its metabolic activation to semiquinone and quinoidal product, *Free Radical Biology and Medicine,* Vol. 26, pp. 107–116, ISSN: 0891-5849.

Minor, W., Steczko, J., Stec, B., Otwinowski, Z., Bolin, J.T., Walter, R. & Axelrod, B. (1996). Crystal structure of soybean lipoxygenase L-1 at 1.4 Å resolution. *Biochemistry,* Vol. 35 pp. 10687–10701.

Mochizuki, N. & Kwon, Y.G. (2008). 15-Lipoxygenase-1 in the vasculature: Expanding roles in angiogenesis, *Circulation Research,* Vol.1 02, pp. 143-145.

Moreno, J.J. (2009). New aspects of the role of hydroxyeicosatetraenoic acids in cell growth and cancer development. *Biochemical Pharmacology,* Vol. 77, pp. 1–10, ISSN: 0006-2952.

Oldham, M.L., Brash, A.R. & Newcomer, M.E. (2005). Insights from the X-ray crystal structure of coral 8R-lipoxygenase: calcium activation via a C2-like domain and a structural basis of product chirality. *The Journal of Biological Chemistry,* Vol. 280, pp. 39545-39552, ISSN: 1083-351X.

Perkampus, H-H. (1992). UV-Vis spectroscopy and its applications. Springer laboratory. Springer Lab Manuals Series. Springer-Verlag. ISBN 3540554211, 9783540554219.

Pidgeon, G.P., Lysaght, J., Krishnamoorthy, S., Reynolds, J.V., O'Byrne, K., Nie, D. & Honn, K.V. (2007). Lipoxygenase metabolism: roles in tumor progression and survival. *Cancer and Metastasis Reviews,* Vol. 26, pp. 503–524, ISSN: 0167-7659.

Pinto, M.C. & Macias, P. (2005). Oxidation of dietary polyphenolics by hydroperoxidase activity of lipoxygenase. *Journal Agricultural Food Chemistry,* Vol. 53, pp. 9225-9230.

Pinto, M.C., Duque, A.L. & Macías, P. (2011). Fluorescence quenching study on the interaction between quercetin and lipoxygenase, *Journal of Fluorescence,* Vol. 21, pp. 1311-1318.

Prasad, N. S., Raghavendra, R., Lokesh, B.R., & Naidu, K. A. (2004). Spice phenolics inhibit human PMNL 5-lipoxygenase. *Prostaglandins, Leukotrienes and Essential Fatty Acids,* Vol. 70, pp. 521-528.

Ramadoss, C.S., Pistorius, E.K. & Axelrod, B. (1978). Coupled oxidation of carotene by lipoxygenase requires two isoenzymes. *Archives of Biochemistry and Biophysics,* Vol. 190, pp. 549–552.

Redrejo-Rodriguez, M., Tejeda-Cano, A., Pinto, M.C. & Macias, P. (2004). Lipoxygenase inhibition by flavonoids: semiempirical study of the structure activity relation. *Journal of Molecular Structure: THEOCHEM,* Vol. 674, pp. 121-124.

Rogers, A. & Gibon, Y. (2009). Enzyme Kinetics: Theory and Practice. In J. Schwender (ed.), *Plant Metabolic Networks,* DOI 10.1007/978-0-387-78745-9_4, Springer Science+Business Media, LLC 2009, URL: http://www.bnl.gov/pubweb/alistairrogers/linkable_files/pdf/Rogers_&_Gibon _2009.pdf.

Sadik, C.D., Sies, H. & Schewe, T. (2003). Inhibition of 15-lipoxygenases by flavonoids: structure-activity relations and mode of action. *Biochemical Pharmacology,* Vol. 65, pp.773-781.

Samuelsson, B., Borgeat, P., Hammarstrom, S. & Murphy, R.C. (1980). Leukotrienes: a new group of biologically active compounds. *Advanced Prostaglandin and Thromboxane Research,* Vol. 6, pp. 1–18.

Santano[a], E., Pinto, M.C., & Macias, P. (2002). Xenobiotic oxidation by hydroperoxidase activity of lipoxygenase immobilized by absorption on controlled pore glass. *Enzyme and Microbial Technology,* Vol. 30, pp. 639-646.

Santano[b], E., Pinto, M.C. & Macias, P. (2002). Chlorpromazine oxidation by hydroperoxidase activity of covalent immobilized lipoxygenase. *Biotechnology and Applied Biochemistry,* Vol. 36, pp. 95-100.

Schewe, T. (2002). 15-lipoxygenase-1: a prooxidant enzyme. *Biological Chem*istry, Vol. 383, pp. 365-374.

Schewe, T. & Sies, H. (2003). Flavonoids as protectants against prooxidant enzymes, *Research monographs*, Institut für Physiologische Chemie I; Heinrich-Heine-Universität Düsseldorf– Date of access: 7.07.2008. Available from: http://www.uni-duesseldorf.de/WWW/MedFak/PhysiolChem/index.html.

Schewe, T., Halangk, W., Hiebsch, C. & Rapoport, S.M. (1975). A lipoxygenase in rabbit reticulocytes which attacks phospholipids and intact mitochondria. *FEBS Letters*, Vol. 60, pp. 149–152.

Schneider, I. & Bucar[a] F. (2005). Lipoxygenase inhibitors from natural plant sources. Part I Medicinal plants with action on arachidonate 5-lipoxygenase. *Phytotherapy Research*, Vol. 19, pp. 81-102.

Schneider, I. & Bucar[b] F. (2005). Lipoxygenase inhibitors from natural plant sources. Part II Medicinal plants with action on arachidonate 12-lipoxygenase, 15-lipoxygenase and leukotriene receptor antagonists. *Phytotherapy Research*, Vol. 19, pp. 263-272.

Schwab, J.M., Chiang, N., Arita, M. & Serhan, C.N. (2007). Resolvin E1 and protectin D1 activate inflammation-resolution programmes. *Nature*, Vol. 447, pp.869–874.

Serhan, C.N., Hamberg, M. & Samuelsson, B. (1984). Lipoxins: novel series of biologically active compounds formed from arachidonic acid in human leukocytes. *Proceedings of the National Academy of Science USA*, Vol. 81, pp. 5335–5339.

Serhan, C.N., Gotlinger, K., Hong, S. & Arita, M. (2004). Resolvins, docosatrienes, and neuroprotectins, novel omega-3-derived mediators, and their aspirin-triggered endogenous epimers: an overview of their protective roles in catabasis. *Prostaglandins and Other Lipid Mediators*, Vol. 73, pp. 155–172.

Serhan, C.N., Yang, R., Martinod, K., Kasuga, K., Pillai, P.S., Porter, T.F., Oh, S.F. & Spite, M. (2009). Maresins: novel macrophage mediators with potent antiinflammatory and proresolving actions. *Journal of Experimental Medicine*, Vol. 206, pp. 15–23.

Siedow, J.N. (1991). Plant lipoxygenases. Structure and function. *Annual Review of Plant Physiology and Plant Molecular Biology*, Vol. 42, pp. 145-188.

Skrzypczak-Jankun E., Amzel L.M., Kroa B.A. & Funk, M.O.Jr. (1997). Structure of soybean lipoxygenase L3 and a comparison with its L1 isoenzyme. *Proteins*, Vol. 29, pp. 15–31.

Skrzypczak-Jankun, E., Bross, R.A., Carroll, R.T., Dunham, W.R. & Funk, M.O. Jr. (2001). Three-dimensional structure of a purple lipoxygenase. *Journal of American Chemical Society*, Vol. 123, pp. 10814–10820.

Skrzypczak-Jankun[a], E., Zhou, K., McCabe, N.P., Selman, S.H. & Jankun. J. (2003). Structure of curcumin in complex with lipoxygenase and its significance in cancer. *International Journal of Molecular Medicine*, Vol. 12, pp.17–24.

Skrzypczak-Jankun[b], E., Zhou, K. & Jankun, J. (2003). Inhibition of lipoxygenase by (2)-epigallo-catechin gallate: X-ray analysis at 2.1 Å reveals degradation of EGCG and shows soybean LOX-3 complex with EGC instead. *International Journal of Molecular Medicine*, Vol. 12, pp. 415–420.

Skrzypczak-Jankun, E., Borbulevych, O.Y. & Jankun, J. (2004). Soybean lipoxygenase-3 in complex with 4-nitrocatechol. *Acta Crystallographica Section D Biological Crystallography*, Vol. 60, pp. 613–615.

Skrzypczak-Jankun, E., Borbulevych, O.Y., Zavodszky, M.I., Baranski, M.R., Padmanabhan, K., Petricek V. & Jankun J. (2006). Effect of crystal freezing and small-molecule binding on internal cavity size in a large protein: X-ray and docking studies of lipoxygenase at ambient and low temperature at 2.0 Å resolution. *Acta Crystallographica Section D Biological Crystallography*, Vol. 62, pp. 766–775.

Szefel, J., Piotrowska, M., Kruszewski, W.J., Jankun, J., Lysiak-Szydlowska, W. & Skrzypczak-Jankun, E. (2011). Eicosanoids in prevention and management of diseases, *Current Molecular Medicine,* Vol. 11, pp. 13-25.

Takahama, U. (1985). O_2-dependent and -independent photooxidation of quercetin in the presence and absence of riboflavin and effects of ascorbate on the photooxidation. *Photochemistry and Photobiology*, Vol. 42(1), pp. 89-91.

Tappel, A. L. (1962). Lipoxidase. In: *Methods in Enzymology*, S. P. Colowick and N. O. Kaplan, (Eds.) Vol 5, pp 539-542, Academic Press, New York and London.

Tappel, A. L., Boyer, P. D. & Lundberg W. O. (1952). The reaction mechanism of soy bean lipoxygenase, *The Journal of Biological Chemistry*, Vol. 199, pp. 267-281, ISSN: 1083-351X.

Theorell, H., Bergstrom, S. & Akeson, A. (1944). On the lipoxidase enzymes in soybean. *Arkiv. Kemi Mineralogi Och Geologi*, 19A, 6.

Theorell, H., Holman, R. T. & Åkeson, Å. (1947). Cristalline lipoxidase. *Acta Chemica Scandinavica*, Vol. 1, pp. 571-576.

Vicaş, S., Chedea, V.S., Dulf, F.V. & Socaciu C. (2006). The influence of quercetin concentration on its intermolecular interaction with soybean Lipoxygenase-1, *Buletinul USAMV-CN*, Vol. 62, pp. 377-380, ISSN 1454-2382.

Zheng, Y. & Brash, A.R. (2010). On the role of molecular oxygen in lipoxygenase activation: comparison and contrast of epidermal lipoxygenase-3 with soybean lipoxygenase-1. *The Journal of Biological Chemistry*, Vol. 285, pp. 39876-3987.

Zsila, F., Bikadi, Z. & Simonyi M. (2003). Probing the binding of the flavonoid, quercetin to human serum albumin by circular dichroism, electronic absorption spectroscopy and molecular modelling methods. *Biochemical Pharmacology*, Vol. 65, pp. 447-456.

http://wiz2.pharm.wayne.edu/biochem/enz.html. Date of access: 5.01.2007.

http://www.wiley.com/college/pratt/0471393878/student/animations/enzyme_inhibitio n/index.html. Date of access: 14.08.2011.

6

Biochemical and Histopathological Toxicity by Multiple Drug Administration

Zeeshan Feroz[1] and Rafeeq Alam Khan[2*]
[1]Ziauddin College of Pharmacy, Ziauddin University, Karachi,
[2]Department of Basic Medical Sciences,
King Saud Bin Abdul Aziz University of Health Sciences, Jeddah,
[1]Pakistan
[2]Kingdom of Saudi Arabia

1. Introduction

With the increase in ways and means to improve health care there has been an increase in miseries of humanity a patient is often presented with several pathological situations that greatly necessitate the need for multiple drug administration and this in turn increases the chances of drug toxicity. Hence there is an immense need to explore such drug combinations that could be given safely to patients with multiple disorders.

Multiple drug administration increases chances of drug interaction, altering the responses of drugs either increased or decreased pharmacological effects, or a new pharmacological response. Generally drug interactions should be avoided, due to the possibility of poor or unexpected outcomes and can be prevented with access to current, comprehensive and reliable information which may improve the safe and cost-effective patient care. Most countries face an augmented load of cardiovascular diseases (CVD) and epilepsy along with chronic non-communicable disorders such as diabetes mellitus. Hence it is essential to recognize the probable toxicities that might occur due to multiple drug administration. Occasionally these toxicities are predictable on the basis of known pharmacology of the drugs used, thus combinations require separate investigations with animal toxicity studies.

1.1 Epilepsy

Epilepsy is the leading neurological disorder in the world categorized by abnormal hyper excitability of the neurons causing seizures with or without loss of consciousness. These seizures are of short-term and an indication of unusual, extreme or synchronous neuronal activity in the brain (Fisher et al., 2005).

Epilepsy represents the 3rd most common neurologic disorder in developed countries after stroke and dementias, encountered in elderly (Lim, 2004). The prevalence of epilepsy is around 0.4 to 0.8 % (Brodie and Dichter, 1996) and its overall occurrence is around 50-70 cases per 100, 000 in developed countries and 100 per 100, 000 in developing countries (Lim, 2004). Epilepsy

---

* Corresponding Author

is a significant, but often underappreciated, health problem in Asia (Mac et al., 2007). It is estimated that approximately 50 million people worldwide have epilepsy (Kwan and Brodie, 2000). This figure had recently reduced and it has been estimated that approximately 45 million of population globally have epilepsy (French and Pedley, 2008). In Pakistan its predominance is approximately 1 % (Aziz et al., 1994 and 1997; Khatri et al., 2003). The utmost occurrence rate of 1.25% was found at the age group 20-29 years. The incidence rate gradually dropped, reaching the lowest of 0.49% in the age group of 50-59. Conversely the prevalence rate augmented again reaching to 1.1% at age > 60 years (Aziz et al., 1994).

Etiology of epilepsy is age related, in children, approximately 20% are remote symptomatic, 50% are cryptogenic while 30% are idiopathic. On the other hand, in elderly, approximately 55% are remote symptomatic whereas 45% are idiopathic/ cryptogenic. In elderly causes and risk factors for seizures are significantly variable (LIM, 2004). The cumulative lifetime risks for epilepsy and unprovoked seizure in industrialized countries are 3.1% and 4.1%, respectively (McHugh and Norman, 2008). In most of the cases (62%) the reason is unidentified, stroke (9.0%), head trauma (9.0%), alcohol (6.0%), neurodegenerative disease (4.0%), static encephalopathy (3.5%), brain tumors (3.0%), and infection (2.0%) account for remaining cases. Although cerebrovascular reasons are more widespread in the old age, the reason is yet to be explored in 25% to 40% of patients who are 65 years of age or older (French and Pedley, 2008). Majority of the patients are well controlled on a single antiepileptic drug (Nadkarni et al., 2005). Since the early 1990s, a number of latest antiepileptic drugs have arrived in the market that proposed considerable benefits in terms of their favorable pharmacokinetics, enhanced tolerability and decrease probability for drug-drug interactions (Bialer and White, 2010). Every new drug offers a special profile of pharmacokinetics, undesirable effects, and mechanisms of action, making best utilization of these agents even more difficult (LaRoche and Helmers, 2004). Seizures can be managed in patients with epilepsy by means of conventional antiepileptic drugs, but regardless of optimal therapy 25% to 30% of patients continue to have seizures and others have undesirable side effects. Hence there is a need for additional drugs as well as new approaches for preventing epilepsy (Dichter and Brodie, 1996; Bialer and White, 2010).

1.2 Hypertension

Hypertension is a widespread human disease that badly affects approximately 1 billion people globally (Varon and Marik, 2003; Dickson and Sigmond, 2006; Chobanian, 2008). Unfortunately, regardless of current progresses in understanding and treating hypertension, its occurrence keeps on increasing. Worldwide 26% of the adult population suffers from hypertension (Dickson and Sigmond, 2006) and its occurrence is projected to boost up to 60% by 2025, when a total of 1.56 billion people may be exaggerated. Pakistan ranks at number sixth in terms of its population (165million in 2007) which is constantly increasing at a rate about 1.83% per year, national heath surveys reveals that 33% of Pakistan's population beyond the age of 45 has hypertension (Wasay and Jabbar, 2009). Hypertension is the foremost threat for cardiovascular disease (Lardinois, 1995; Peralta et al., 2007; Chobanian, 2008) responsible for one half of the coronary heart disease and about two third of the cerebrovascular disease load (Cutler et al., 2008) and is accountable for the most of deaths globally (Adrogue and Madias, 2007), even moderate increase in arterial blood pressure results in reduced life span.

1.3 Diabetes

Diabetes mellitus and hypertension are widespread that exist together at a larger rate than the individual one (Sowers and Zemel, 1990; Epstein and Sowers, 1992; Tenenbaum et al., 1999; Zanella et al., 2001). The occurrence of hypertension in the diabetic individual noticeably enhances the threat and hastens the course of cardiac disease, peripheral vascular disease, stroke, retinopathy, and nephropathy (Epstein and Sowers, 1992; Zanella et al., 2001). The occurrence of simultaneous hypertension and diabetes appears to be growing in developed nations because populations are aging and both hypertension and non-insulin dependent diabetes mellitus occurrence increases as the age progresses (Sowers and Epstein, 1995). People with diabetes faces two to four times augmented risk of CVD in contrast to the general population, simultaneous hypertension triples the already high risk of coronary artery disease, doubles total mortality and stroke risk, and may be accountable for up to 75% of all CVD in people with diabetes (Stults and Jones, 2006).

Rates of diabetes are increasing around the world (Kassab et al., 2001) which now becomes one of the major public health challenges for the 21st century. The increase occurrence in diabetes is because of aging population, obesity and stressing life style. Poverty has been under recognized as a contributor to prevalence of diabetes but it is strongly associated with the unhealthy alimentary habits (Krier et al., 1999; Riste et al., 2001). The incidence of diabetes get higher in the last decade because of factors that are strongly related to the life style as is inactivity and population aging (Muchmore et al., 1994; Keen, 1998). Studies show that type 2 diabetes affects 3% to 5% of the population in some countries and type 1 moves towards the younger ages (Dixon, 2002; Petkova et al., 2006).

The increase rate of diabetes will be noticeably higher in developing countries, between 1995 and 2025, the number of persons with diabetes is predictable to enhance by 170% in the developing world, in contrast with 42% in developed nations. Hence, by the year 2025 above 75% of the people with diabetes will exist in developing countries (Nicolucci et al., 2006). A national health survey of Pakistan reveals that 25% of patients above 45 years have diabetes mellitus and Pakistan ranks number six globally in terms of prevalence of diabetes. It was projected that in 2000 there were 5.2 million diabetic patients and this will increase to 13.9 million by 2020, leading Pakistan to 4th most populous country for diabetes mellitus (Wasay and Jabbar, 2009). However a survey conducted in 2010 by Hayat and Shaikh reveals that Pakistan ranks number seven on diabetes prevalence and figures show that about 6.9 million people have diabetes. The International Diabetes Federation predicts that this number will rise to 11.5 million by 2025 if effective procedures are not taken to control the disease (Jawad, 2003).

The occurrence of diabetes is continually growing and rising at a distressing rate, and it is projected that, unless successful prevention and control measures are put into practice, this disease will soon involve 300 million persons worldwide (Sowers, 2004). Globally more than 170 million people have diabetes, and this figure is expected to be more than double by the year 2030, if existing trends continue (Boden and Taggart, 2009, Hoque et al., 2009).

1.4 Arrhythmia

Hypertension is usually linked with arrhythmias in patients with and without simultaneous CVD. There are studies which show the possible links between hypertension and atrial and ventricular arrhythmias, though the principal pathophysiological mechanism remains unclear (Yiu and Tse, 2008). The prevalence and risk factors for arrhythmias vary among men and women (Wolbrette et al., 2002). The most prevalent arrhythmia seen in clinical practice is atrial fibrillation which currently influences more than 2 million Americans, with an expected rise to 10 million by the year 2050 (Zimetbaum, 2007).

A patient is often presented with several other pathological states along with epilepsy; such as hypertension, arrhythmias, and diabetes, therefore it is essential to discover the drug-drug interaction upon simultaneous use of anti-epileptic with antihypertensive, antiarrhythmic and antidiabetic. A well reported example is the increase in serum phenytoin levels when used concomitantly with amiodarone and therefore resulting in phenytoin toxicity (Lesko, 1989; Nolan et al., 1990) thus there is an massive need to assess the toxicities of multiple drug administration and to explore relatively safe combination for individuals with multiple disorders, not to predict but rather to warn the users and prescribers, of the possible dangers, to discourage the use of combination which have high cumulative toxicities in animals and to suggest more useful combination in countries where drug regulatory control is very poor.

2. Biochemical testing and histopathological examination of liver toxicities

Serum biochemical parameters can provide important and useful information in assessing not only the extent and severity of liver damage, but also the type of liver damage (Ramaiah, 2007). Histopathological assessments also take part in the diagnosis of liver disease; moreover evaluation of morphological changes may provide additional information that may be useful for clinical management for example, grading of inflammatory activity and staging of fibrosis in chronic viral hepatitis, and the distinction between simple steatosis and steatohepatitis in alcoholic and non-alcoholic fatty liver disease (Hubscher, 2006).

Liver function tests (LFT) are helpful screening tools to detect hepatic dysfunction (Kim, 2008; Thapa and Walia, 2007; Astegiano et al., 2004). Since liver performs a variety of functions, no single test is sufficient enough to provide complete estimate of liver functions (Kim, 2008; Astegiano et al., 2004).

Table 1A and 1B reveals the comparison of γ–glutamyl transferase (γ-GT), alkaline phosphatase (ALP), alanine transaminases (ALT), total bilirubin (TBR) and direct bilirubin (DBR) levels between control animals and animals kept on individual drugs and their combinations for a period of 60 days and then after drug free interval of 15 days in normal therapeutic doses. The administration of amiodarone (4.285 mg/kg) in rabbits shows highly significant elevation in the levels of serum γ–GT, ALP, ALT and DBR (Feroz et al., 2011a). There are studies in which long-term administration of amiodarone was associated with fatal hepatotoxicity (Richer and Robert, 1995; Usdin et al., 1996; Mendez et al., 1999) although most hepatic adverse effects were transient and reversible; however deaths have also been reported from amiodarone-induced hepatotoxicity (Richer and Robert, 1995). Microscopic examination of the hepatic tissue has shown mild diffuse cellular swelling in hepatocytes (Fig 1B). Moreover the administration of losartan potassium (0.892 mg/kg) and verapamil (1.714 mg/kg) revealed highly significant elevation in serum ALP, elevations in serum ALP initiate predominantly from liver and bone (Renner and Dallenbach, 1992). There was also a significant elevation in TBR in animals kept on verapamil alone (Feroz et al., 2011a). However no significant changes have been found in animals kept on glibenclamide (0.125 mg/kg), oxcarbazepine (18.5 mg/kg) and captopril (0.512 mg/Kg) alone.

Data from animal's studies also shows that the administration of amiodarone-glibenclamide-losartanpotassium- oxcarbazepine (AGLO) combination causes highly significant elevation in serum ALP. The transaminases, ALP and γ -GT are most widely used as indicators of hepatobiliary disease (Renner and Dallenbach, 1992). Increase in ALP might be due to cholestasis (Giannini et al., 2005) or it may also suggest a biliary tract disorder (Herlong, 1994). This explains that rise in ALP might be due to chloestatic diseases or partial obstruction of bile ducts or primary biliary cirrhosis (Pratt and Kaplan, 2000). Microscopic examination of hepatic tissue has shown moderate portal inflammation (Fig 1C). There was also a significant decreased in TBR in animals kept on AGLO combination (Feroz et al., 2011a).

The administration of amiodarone-glibenclamide-verapamil-oxcarbazepine (AGVO) combination in animals shows highly significant elevation in γ -GT. Its level is elevated in a number of pathological conditions such as pancreatic disease, myocardial infarction, renal failure, chronic obstructive pulmonary disease, diabetes, and alcoholism. Measurement of serum γ -GT offers the presence or absence of hepatobiliary disease. Microscopic examination of hepatic tissue has shown congestion and mild mononuclear inflammatory infiltrate (Fig 1D). There was also a significant decreased in TBR in animals kept on AGVO combination (Feroz et al., 2011a). There has been a highly significant and significant decreased in ALT and TBR in animals kept on amiodarone-glibenclamide-captopril-oxcarbazepine (AGCO) combination (Feroz et al., 2011a). Aminotransferase levels are sensitive indicators of liver-cell injury and are useful in identifying the hepatocellular disease (Pratt and Kaplan, 2000). Abnormal AST and ALT point to a hepatocyte disorder (Herlong, 1994). However microscopic examination shows congestion only (Fig 1E) illustrating no remarkable changes in the hepatic tissue of these animals (Feroz et al., 2011a).

Parameters/ Groups	γ-GT (μ/l)	ALP (μ/l)	ALT (μ/l)	TBR (mg/dl)	DBR (mg/dl)
Control	11.28+1.21	55.51+5.92	83.16+5.21	0.32+0.03	0.15+0.01
Amiodarone	19.41+1.43**	88.34+5.64**	96.60+1.52*	0.30+0.02	0.32+0.05**
Glibenclamide	11.02+0.75	62.82+2.06	68.50+4.72	0.23+0.03	0.20+0.02
Los. Pot	8.82+0.80	79.0+3.25**	81.40+7.98	0.21+0.05	0.14+0.02
Oxcarbazepine	14.46+1.41	53.47+1.26	79.26+2.11	0.25+0.01	0.17+0.02
Verapamil	10.77+1.37	80.44+6.89**	80.07+3.67	0.53+0.12*	0.13+0.01
Captopril	13.58+0.55	57.37+2.10	73.94+3.31	0.32+0.06	0.15+0.01
AGLO	11.21+1.20	79.21+5.11**	76.55+4.16	0.13+0.01*	0.18+0.03
AGVO	16.78+0.95**	66.62+7.77	70.76+4.25	0.13+0.03*	0.21+0.04
AGCO	11.95+1.48	47.75+4.03	56.74+4.69**	0.14+0.04*	0.21+0.04

n=9

Mean ± S.E.M

*p < 0.05 significant with respect to control

**p <0.005 highly significant with respect to control

Table 1A.Comparison of hepatic parameters following 60 days administration of individual drugs and their combinations {Adopted from Feroz et al., 2011(a)}.

Parameters/ Groups	γ-GT (μ/l)	ALP (μ/l)	ALT (μ/l)	TBR (mg/dl)	DBR (mg/dl)
Control	11.30±1.21	55.30±5.90	82.70±5.20	0.31±0.03	0.15±0.01
Amiodarone	19.16±1.42**	88.0±5.70**	96.56±1.50*	0.26±0.02	0.30±0.04**
Glibenclamide	10.58±0.82	62.42±2.10	67.80±4.80	0.22±0.03	0.16±0.01
Los. Pot	8.51± 0.78	79.63±3.10**	80.80±8.0	0.17±0.05	0.13±0.02
Oxcarbazepine	14.17±1.40	53.96±1.30	78.72±2.11	0.23±0.01	0.16±0.01
Verapamil	10.39±1.30	81.0±6.80**	79.4±3.60	0.50±0.12*	0.12±0.01
Captopril	13.17±0.51	56.93±2.20	73.39±3.30	0.30±0.06	0.14±0.01
AGLO	10.79±1.10	79.0±5.10**	76.10±4.20	0.12±0.01*	0.17±0.03
AGVO	16.42±0.92**	66.0±7.70	70.51±4.22	0.11±0.02*	0.18±0.03
AGCO	11.43±1.50	46.90±4.0	54.0±5.10**	0.13±0.04*	0.18±0.03

n=9

Mean ± S.E.M

*p < 0.05 significant with respect to control

**p <0.005 highly significant with respect to control

Table 1B. Comparison of hepatic parameters following drug-free interval of 15 days of individual drugs and their combinations {Adopted from Feroz et al., 2011(a)}.

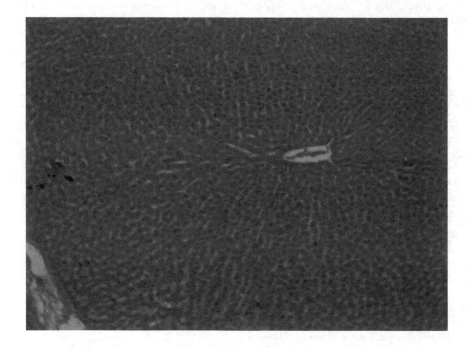

Fig. 1A. Hepatic tissue showing no microscopic change {Adopted from Feroz et al., 2011(a)}.

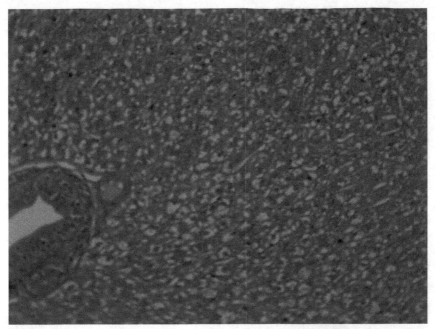

Fig. 1B. Hepatic tissue showing cellular swelling {Adopted from Feroz et al., 2011(a)}.

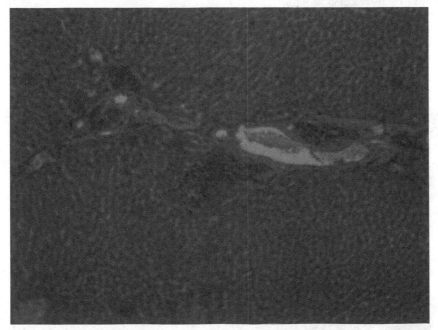

Fig. 1C. Hepatic tissue showing moderate portal inflammation {Adopted from Feroz et al., 2011(a)}.

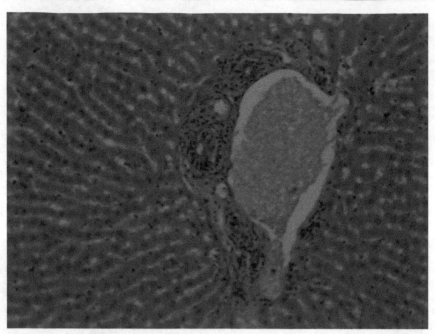

Fig. 1D. Hepatic tissue showing congestion and mild mononuclear inflammatory infiltrate {Adopted from Feroz et al., 2011(a)}.

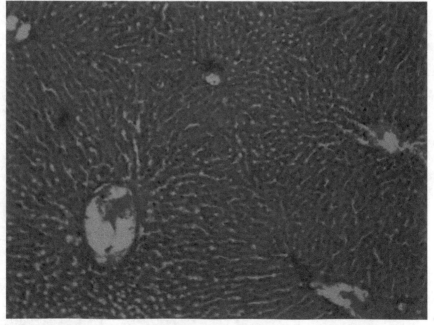

Fig. 1E. Hepatic tissue showing congestion {Adopted from Feroz et al., 2011(a)}.

3. Biochemical testing and histopathological examination of renal toxicities

Drugs are a frequent cause of acute kidney injury particularly in patients older than 60 years and have a higher occurrence of diabetes and CVD (Naughton, 2008). Renal function test are important in assessing the amount and severity of renal damage.

Table 2A and 2B reveals the comparison of urea and creatinine levels between control animals and animals kept on individual drugs and their combinations for a period of 60 days and then after drug free interval of 15 days in normal therapeutic doses. The administration of amiodarone in animals causes increase in the levels of urea and creatinine, whereas animals received verapamil showed significant increase in the level of urea only. However biochemical changes do not correlate to histopathological changes in renal tissue (Fig 2A) hence it is not an indication of renal damage (Feroz et al., 2010). Animals received glibenclamide showed no significant changes at biochemical level; however microscopic examination of renal tissue reveals mild tubulointerestial nephritis (Fig 2B), which together with insignificant rise in serum urea might be an indicative of developing renal damage. Significant elevation in serum urea level in animals received amiodarone and verapamil alone after drug free interval might be due to slow excretion rate of amiodarone and verapamil from the body.

Animals received AGVO combination showed highly significant rise in urea level though it was reversed after the drug-free interval, while animals kept on AGCO combination showed highly significant elevation of serum urea and creatinine. Increased creatinine level suggests decreased creatinine clearance which is a reliable indicator of decreased glomerular filtration rate due to renal damage. However after dug-free interval urea level remained highly significant while creatinine level was changed from highly significant to significant (Feroz et al., 2010).

Parameters/ Groups	Urea (mg/dl)	Creatinine (mg/dl)
Control	52.12+4.37	0.64+0.02
Amiodarone	169.65+5.73**	1.12+0.07**
Glibenclamide	52.80+2.61	0.70+0.05
Los. Pot	55.05+1.50	0.64+0.02
Oxcarbazepine	49.27+1.40	0.58+0.07
Verapamil	68.14+1.62*	0.71+0.02
Captopril	62.0+1.83	0.60+0.02
AGLO	55.28+2.44	0.64+0.07
AGVO	74.86+6.57**	0.71+0.05
AGCO	148.54+6.81**	1.20+0.05**

n=9

Mean + S.E.M

*p < 0.05 significant with respect to control

**p <0.005 highly significant with respect to control

Table 2A. Comparison of renal parameters following 60 days administration of individual drugs and their combinations (Adopted from Feroz et al., 2010).

Parameters/ Groups	Urea (mg/dl)	Creatinine (mg/dl)
Control	51.97+4.37	0.63+0.02
Amiodarone	68.15+1.30*	0.70+0.03
Glibenclamide	52.40+2.59	0.66+0.03
Los. Pot	54.61+1.41	0.73+0.05
Oxcarbazepine	49.0+1.41	0.54+0.03
Verapamil	61.88+1.46*	0.64+0.03
Captopril	60.77+1.57	0.67+0.02
AGLO	54.66+2.45	0.68+0.07
AGVO	59.20+4.48	0.63+0.05
AGCO	120.65+6.42**	0.75+0.03*

n=9

Mean $\pm$ S.E.M

*p < 0.05 significant with respect to control

**p <0.005 highly significant with respect to control

Table 2B. Comparison of renal parameters following drug-free interval of 15 days of individual drugs and their combinations (Adopted from Feroz et al., 2010).

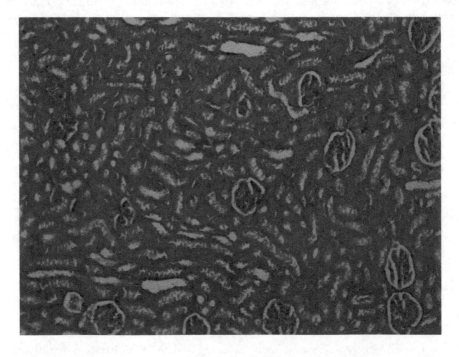

Fig. 2A. Renal tissue showing no microscopic change (Adopted from Feroz et al., 2010).

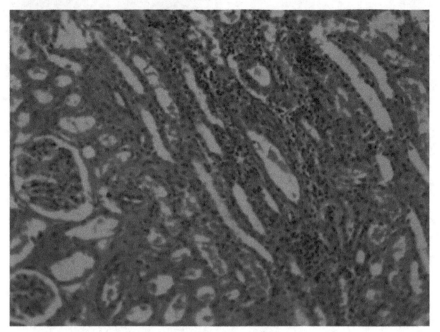

Fig. 2B. Renal tissue showing mild tubulointerestial nephritis (Adopted from Feroz et al., 2010).

4. Biochemical testing and histopathological examination of cardiac toxicities

Cardiac enzymes are proteins that escape out of injured myocardial cells resulting in elevated levels in blood. Table 3A and 3B reveals the comparison of creatinine kinase (CK) and aspartate transaminases (AST) levels between control animals and animals kept on individual drugs and their combinations for a period of 60 days and then after drug free interval of 15 days in normal therapeutic doses. Animals received amiodarone, losartan potassium, oxcarbazepine and captopril alone revealed highly significant elevation in the level of CK but these changes do not correlate with histological changes (Fig 3A) (Feroz et al., 2010). However animals kept on oxcarbazepine alone revealed significant elevation in CK level even after drug free interval which might be an indication of developing neuroleptic malignant syndrome (Pelonero et al., 1998).

The administration of AGLO combination in animals causes significant elevation in CK and AST levels, moreover inflammatory changes in the cardiac tissues (Fig 3B) suggest possible cardiac injury (Feroz et al., 2010). There are studies which suggest that raise in CK level increases the risk of myocardial infarction (Smith et al., 1976; Kumar et al., 2003; Watanabe et al., 2009). Thus simultaneous elevation of AST along with CK and histological changes might be indicative of severe myocardial cellular damage (Kratz et al., 2002). There was significant decrease in CK level in animal received AGCO combination at the end of dosing and after drug-free interval which may be due to reduced muscle mass, wasting or cachexia (Feroz et al., 2010).

Parameters/ Groups	CK (µ/l)	AST (µ/l)
Control	311.35+12.57	54.07+1.03
Amiodarone	567.42+29.61**	54.81+2.37
Glibenclamide	321.26+13.48	48.63+4.69
Los. Pot	400.84+4.24**	51.36+0.99
Oxcarbazepine	392.61+14.86**	56.05+1.99
Verapamil	307.28+6.49	49.42+3.01
Captopril	590.12+12.59**	57.13+0.80
AGLO	393.91+21.54**	60.66+1.93*
AGVO	330.11+38.41	53.05+1.48
AGCO	275.40+7.30*	50.87+1.29

n=9

Mean ± S.E.M

*p < 0.05 significant with respect to control

**p <0.005 highly significant with respect to control

Table 3A. Comparison of cardiac parameters following 60 days administration of individual drugs and their combinations (Adopted from Feroz et al., 2010).

Parameters/ Groups	CK (µ/l)	AST (µ/l)
Control	311.01+12.59	53.78+1.02
Amiodarone	304.16+9.45	54.63+2.32
Glibenclamide	319.67+12.81	53.24+5.45
Los. Pot	310.81+3.57	51.33+0.94
Oxcarbazepine	354.12+15.38*	56.03+1.93
Verapamil	305.35+6.63	49.28+2.96
Captopril	346.63+17.37	56.42+0.73
AGLO	270.56+13.33*	56.57+0.88
AGVO	307.34+23.83	53.36+1.65
AGCO	271.31+6.70*	50.75+1.29

n=9

Mean ± S.E.M

*p < 0.05 significant with respect to control

**p <0.005 highly significant with respect to control

Table 3B. Comparison of cardiac parameters following drug-free interval of 15 days of individual drugs and their combinations (Adopted from Feroz et al., 2010).

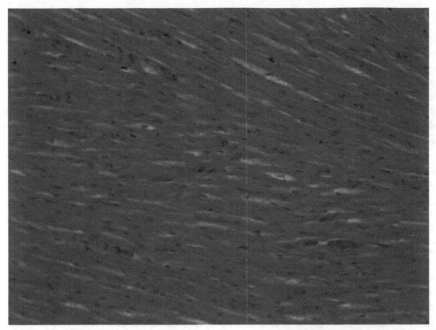

Fig. 3A. Cardiac tissue showing no microscopic change (Adopted from Feroz et al., 2010).

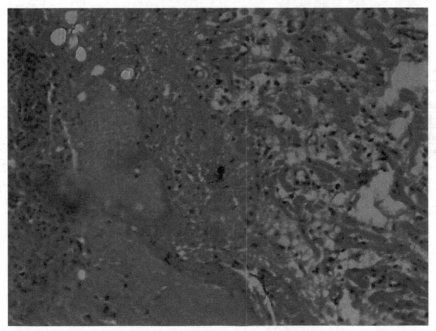

Fig. 3B. Cardiac tissue showing focal pericardial inflammation (Adopted from Feroz et al., 2010).

5. Biochemical testing of lipid profile

Cholesterol and triglycerides are the most important plasma lipids, crucial for formation of cell membrane, synthesis of hormones and offer a source of free fatty acids (Dietschy, 1998). Table 4A and 4B reveals the comparison of cholesterol, triglyceride, HDL-C (high density lipoprotein cholesterol) and LDL-C (low density lipoprotein cholesterol) levels between control animals and animals kept on individual drugs and their combinations for a period of 60 days and then after drug free interval of 15 days in normal therapeutic doses. The administration of losartan potassium and captopril in animals showed highly significant decrease in the level of triglyceride, whereas animals kept on AGLO and AGCO combinations showed highly significant increase in cholesterol at the end of dosing which remained significant even after drug-free interval. There was also highly significant increase in LDL-C level in animals kept on AGLO and AGCO combinations which remained significant even after drug-free interval (Feroz et al., 2011b). Elevated levels of cholesterol and LDL-C are undoubtedly associated with enhanced threat of coronary heart disease (Brown, 1984) and cerebrovascular morbidity and mortality. There has been a correlation among increased LDL-C and atherosclerosis. Since LDL-C gets deposited in the walls of the blood vessel forming atherosclerotic plaque. There are studies which recommend that pathological process could be inverted by dropping the serum LDL-C level (Ross, 1993). There was also significant increase in HDL-C in animals kept on oxcarbazepine alone and AGCO in combination at the end of dosing and following drug-free interval; however reason of elevated HDL-C is yet to be explored (Feroz et al., 2011b).

Parameters/ Groups	Cholesterol (mg/dl)	Triglyceride (mg/dl)	HDL-C (mg/dl)	LDL-C (mg/dl)
Control	91.84+2.65	102.65+2.45	3.10+0.13	28.15+1.90
Amiodarone	88.80+0.62	100.51+2.97	3.07+0.04	27.04+1.13
Glibenclamide	93.37+3.07	106.77+4.71	3.28+0.06	23.25+2.08
Los. Pot	92.77+0.57	91.94+2.86*	3.15+0.07	34.87+1.70
Oxcarbazepine	95.08+1.26	97.08+1.45	3.37+0.05*	29.38+2.03
Verapamil	96.90+1.13	108.53+3.32	3.31+0.08	25.24+1.95
Captopril	96.06+2.89	93.40+4.47*	3.32+0.07	34.18+2.26
AGLO	105.15+3.94**	100.20+2.90	2.93+0.05	46.35+3.50**
AGVO	95.93+1.21	104.46+1.31	3.27+0.04	27.41+1.80
AGCO	173.53+4.22**	95.35+3.17	3.40+0.08**	108.83+6.13**

n=9

Mean ± S.E.M

*p < 0.05 significant with respect to control

**p <0.005 highly significant with respect to control

Table 4A. Comparison of lipid profile following 60 days administration of individual drugs and their combinations {Adopted from Feroz et al., 2011(b)}.

Parameters/ Groups	Cholesterol (mg/dl)	Triglyceride (mg/dl)	HDL-C (mg/dl)	LDL-C (mg/dl)
Control	91.61+2.52	102.12+2.46	2.98+0.12	30.56+1.68
Amiodarone	88.18+0.75	100.26+2.93	2.96+0.04	28.81+1.25
Glibenclamide	93.80+3.10	106.7+4.59	3.16+0.05	26.38+2.06
Los. Pot	92.66+0.65	95.72+1.94	3.04+0.06	34.43+1.22
Oxcarbazepine	94.82+1.33	96.91+1.52	3.26+0.05**	31.38+1.96
Verapamil	96.01+1.27	104.27+1.81	3.18+0.07	29.60+1.90
Captopril	95.37+2.87	95.64+3.69	3.21+0.07	36.05+1.65
AGLO	104.31+3.91**	97.97+2.48	2.83+0.05	48.71+3.31**
AGVO	94.98+1.33	102.40+2.11	3.13+0.04	30.85+1.62
AGCO	99.73+1.18*	94.72+3.13	3.26+0.05**	37.88+2.70**

n=9
Mean $\pm$ S.E.M
*$p < 0.05$ significant with respect to control
**$p < 0.005$ highly significant with respect to control

Table 4B. Comparison of lipid profile following drug-free interval of 15 days of individual drugs and their combinations {Adopted from Feroz et al., 2011(b)}.

6. Biochemical testing of glucose

Table 5A and 5B reveals the comparison of glucose level between control animals and animals kept on individual drugs and their combinations for a period of 60 days and then after drug free interval of 15 days in normal therapeutic doses. AGCO combination showed significant increase in glucose level in rabbits at the completion of dosing period of 60 days and following drug-free interval (Feroz et al., 2011b). Elevated blood glucose level may be due to elevation in the level of cholesterol and LDL-C, because diabetes mellitus is a group of heterogenous, autoimmune, hormonal and metabolic disorders, frequently occurs along with hypertension, hyperlipidemia and obesity (Mahomed and Ojewole, 2003), which also augmented the possibility of coronary heart disease (Howard et al., 2000), however threat of CVD fatality in diabetic persons may be as high as that in non-diabetic persons with prior myocardial infarction (Haffner et al., 1998). There was also a significant elevation in glucose level in animals kept on captopril and oxcarbazepine alone, however it has to be elucidated. Conversely animal kept on glibenclamide alone revealed highly significant decrease in glucose level because the major mechanism of action of glibenclamide is the stimulation of insulin release and the inhibition of glucagon secretion; conversely it was inverted following drug-free interval (Feroz et al., 2011b).

Parameter/ Groups	Glucose (mg/dl)
Control	122.20+7.60
Amiodarone	111.84+3.30
Glibenclamide	75.20+5.79**
Los. Pot	119.67+4.0
Oxcarbazepine	142.17+4.54**
Verapamil	123.33+2.31
Captopril	146.06+4.72**
AGLO	102.50+6.12
AGVO	111.88+3.11
AGCO	145.44+2.93**

n=9
Mean ± S.E.M
*p < 0.05 significant with respect to control
**p <0.005 highly significant with respect to control

Table 5A. Comparison of glucose following 60 days administration of individual drugs and their combinations {Adopted from Feroz et al., 2011(b)}

Parameter/ Groups	Glucose (mg/dl)
Control	123.40+7.40
Amiodarone	113.76+3.10
Glibenclamide	104.50+5.40
Los. Pot	122.30+4.30
Oxcarbazepine	141.50+4.70
Verapamil	124.39+2.30
Captopril	144.30+4.30*
AGLO	102.60+6.10
AGVO	112.11+3.10
AGCO	137.90+3.60*

n=9
Mean ± S.E.M
*p < 0.05 significant with respect to control
**p <0.005 highly significant with respect to control

Table 5B. Comparison of glucose following drug-free interval of 15 days of individual drugs and their combinations {Adopted from Feroz et al., 2011(b)}.

7. Biochemical testing of electrolytes

Table 6A and 6B reveals the comparison of sodium, potassium and calcium concentrations between control animals and animals kept on individual drugs and their combinations for a period of 60 days and then after drug free interval of 15 days in normal therapeutic doses. The balance of electrolytes in our bodies is essential for normal cellular function, since it promotes fluid balance, maintain blood volume, facilitate fluid absorption and generate impulses. Significant alterations may occur in electrolytes following multiple drug administration. There has been significant decrease in concentration of sodium in animals received amiodarone (Feroz et al., 2009), which has potential for significant morbidity and mortality (Goh, 2004). However this decrease became insignificant following drug-free interval. Similarly there was highly significant decrease in serum calcium in animal received glibenclamide, losartan potassium, verapamil, oxcarbazepine, captopril and combination of these drugs (Feroz et al., 2009). Calcium is essentially required for development and maintenance of bones, not only regulate nerve function, but also contributes to the contraction of the muscles and heart. There are studies which suggest that amiodarone induces vitamin D deficiency in individuals not exposed to sunlight (Campbell and Allain, 2006). Vitamin D is essentially required for absorption of calcium; hence in the study by Feroz et al 2009 hypocalcaemia in animals on amiodarone alone or in combination might be due to the deficiency of vitamin D (Cooper and Gittoes, 2008). However reason for hypocalcaemia in other animal groups is yet to be explored. There was highly significant increase in calcium after drug free interval in animals received AGVO combination, this increase in calcium might be due to increase bone resorption, or gastrointestinal absorption or decreased elimination by the kidneys (Strewler, 2000). Hypercalcemia is always a concern, because elevated concentrations can result in renal failure, mineralization of the other soft tissues, cardiac arrhythmia and dysfunction (Sakals et al., 2006). Animals received AGCO combination also showed decrease in potassium level at the end of dosing as well as following drug-free interval. A decreased serum potassium concentration points to disturbance in normal homeostasis which might be an indication of muscle necrosis. However potassium level in animals received AGVO combination was significantly increase after drug-free interval. Hyperkalemia because of drugs most frequently occurs from impaired renal potassium excretion. On the other hand, disturbed cellular uptake of a potassium load as well as unnecessary intake or infusion of potassium-containing substances may also induce hyperkalemia. Therefore prescribing physicians must be conscious about medications that can precipitate hyperkalemia (Perazella, 2000).

Parameters/ Groups	Sodium (µg/ml)	Potassium (µg/ml)	Calcium (µg/ml)
Control	178.50+5.40	5.96+0.38	16.80+1.30
Amiodarone	156.16+3.40*	5.46+0.66	11.34+0.23**
Glibenclamide	187.80+8.90	6.44+0.61	11.04+0.91**
Los. Pot	171.92+2.80	5.36+0.34	11.70+0.33**
Oxcarbazepine	182.70+6.10	5.40+0.24	10.58+0.33**
Verapamil	181.80+9.40	6.04+0.43	11.72+0.52**
Captopril	182.40 11.0	5.50+0.37	10.78+0.52**
AGLO	185.60+11.0	5.26+0.14	10.30+0.51**
AGVO	184.40+5.90	5.66+0.30	12.50+0.59**
AGCO	173.30+7.0	4.54+0.32*	8.92+0.80**

n=5
Mean ± S.E.M
*p < 0.05 significant with respect to control
**p <0.005 highly significant with respect to control

Table 6A. Comparison of sodium, potassium and calcium following 60 days administration of individual drugs and their combinations (Adopted from Feroz et al., 2009).

Parameters/ Groups	Sodium (µg/ml)	Potassium (µg/ml)	Calcium (µg/ml)
Control	178.70±5.80	5.90+0.37	16.80+1.25
Amiodarone	162.72±2.30	5.40±0.65	15.70+1.20
Glibenclamide	181.46±2.90	6.38±0.59	17.04+0.91
Los. Pot	172.50±2.80	5.28±0.31	17.70+0.33
Oxcarbazepine	182.90±6.0	5.26±0.24	16.58+0.33
Verapamil	181.40±9.40	6.0+0.44	17.72+0.52
Captopril	180.60±10.0	5.48±0.33	16.72+0.49
AGLO	175.70±7.40	5.22±0.10	16.30±0.49
AGVO	175.56±3.10	7.0±0.07*	20.38±0.28**
AGCO	170.40±5.90	3.94±0.15**	13.16±0.87**

n=5

Mean ± S.E.M

*$p < 0.05$ significant with respect to control

**$p < 0.005$ highly significant with respect to control

Table 6B. Comparison of sodium, potassium and calcium following drug-free interval of 15 days of individual drugs and their combinations (Adopted from Feroz et al., 2009).

8. Hematological testing

Table 7A and 7B reveals the comparison of hemoglobin concentration, platelet, leucocytes and erythrocytes count between control animals and animals kept on individual drugs and their combinations for a period of 60 days and then after drug free interval of 15 days in normal therapeutic doses. Changes in hematological parameters such as erythrocytes, leucocytes and platelet count and hemoglobin had always a serious concern following administration of drugs individually as well as in combination. There has been significant increase in platelet count in animal group received captopril and oxcarbazepine alone (Feroz et al., 2011a). Increased in platelet might be due to inflammatory disorder or iron deficiency anemia (Schafer, 2004), however there was also a significant increase in leucocytes count in animal group kept on oxcarbazepine alone, on the other hand animal group received amiodarone alone showed significant decrease in leucocytes count which might be due to disturbance in immune system, where as platelet count was not changed significantly, though amiodarone is known to produce thrombocytopenia (Weinberger et al., 1987).

Study conducted by Feroz et al 2011a revealed more severe hematological changes in animals received drugs in combination throughout the experimental period in comparison to animals received the drugs individually. Concurrent administration of AGLO combination showed a significant increased in leucocytes count which might be an indicator of an infection, inflammation, or allergy. On the other hand concurrent administration of AGVO combination showed highly significant increase in erythrocytes count while the other hematological parameters were not altered significantly.

There was significant increase and decrease in leucocytes and platelet count respectively in animals kept on AGCO combination. Decrease in platelet count may be due to insufficient production of platelet in bone marrow, a variety of reasons such as leukemia, lymphomas and several bone marrow disorders may have this effect on platelet count (McMillan, 2007).

Spleen enlargement may also decrease platelet count, or it may probably due to folic acid deficiency (Mant et al., 1979).

Parameters/ Groups	Hemoglobin (mg/dl)	Platelet (x10⁵/c.mm)	Leucocytes (x10³/c.mm)	Erythrocytes (x10⁶/c.mm)
Control	10.62+0.23	412+39	6.26+0.35	6.14+0.16
Amiodarone	11.04+0.37	539+56	3.73+0.44*	6.12+0.26
Glibenclamide	9.90+0.28	433+62	5.61+0.57	5.83+0.19
Los. Pot	10.53+0.72	417+40	5.10+0.69	6.05+0.44
Oxcarbazepine	12.33+1.03	561+50*	8.99+0.78*	6.91+0.67
Verapamil	9.13+0.62	396+48	5.47+0.55	5.25+0.31
Captopril	10.06+0.66	559+66*	4.97+0.63	5.66+0.49
AGLO	10.54+0.69	310+17	8.43+1.24*	5.38+0.23
AGVO	12.01+0.93	320+35	7.74+0.88	9.65+1.22**
AGCO	9.91+0.53	279+16*	8.75+1.01*	6.23+0.36

n=9

Mean ± S.E.M

*p < 0.05 significant with respect to control

**p <0.005 highly significant with respect to control

Table 7A. Comparison of hematological parameters following 60 days administration of individual drugs and their combinations {Adopted from Feroz et al., 2011(a)}.

Parameters/ Groups	Hemoglobin (mg/dl)	Platelet (x10⁵/c.mm)	Leucocytes (x10³/c.mm)	Erythrocytes (x10⁶/c.mm)
Control	10.58+0.24	416+40	6.26+0.36	6.03+0.20
Amiodarone	10.84+0.33	541+55	3.66+0.45*	6.18+0.27
Glibenclamide	9.86+0.25	445+57	5.47+0.59	5.85+0.20
Los. Pot	10.44+0.76	429+37	5.01+0.69	6.09+0.45
Oxcarbazepine	12.23+1.10	549+43*	8.90+0.77*	6.98+0.66
Verapamil	9.03+0.59	385+46	5.43+0.56	5.24+0.31
Captopril	10.02+0.64	558+66*	4.93+0.63	5.66+0.49
AGLO	10.42+0.72	324+21	8.40+1.22*	5.38+0.24
AGVO	12.03+0.87	317+34	7.63+0.87	9.65+1.22*
AGCO	9.73+0.54	281+16*	8.62+0.99*	6.27+0.37

n=9

Mean ± S.E.M

*p < 0.05 significant with respect to control

**p <0.005 highly significant with respect to control

Table 7B. Comparison of hematological parameters following drug-free interval of 15 days of individual drugs and their combinations {Adopted from Feroz et al., 2011(a)}.

9. Conclusion

The problems associated with drug therapy are a significant challenge to health care providers, especially in developing countries where health care system is poor. Minimizing the risk for drug interactions is the desirable aim in drug therapy, since interactions can leads to significant morbidity, mortality and patient quality of life. Individuals taking multiple medications are at increased threat of adverse drug reactions; hence when ever multiple drugs are to be administered in case of multiple disorders such as epilepsy, hypertension, diabetes mellitus and arrhythmias drug treatment should be monitored to avoid adverse effects of the drugs. Studies conducted by Feroz et al not only provides valuable information pertaining to gross toxicities, microscopic changes and toxic effects on hepatic, renal, cardiac, lipid profile, glucose, electrolytes and hematological parameters but also give clues about the drug combination having higher incidence of cumulative toxicities.

These studies in general has revealed that animals received AGCO combination comparatively showed higher toxicities with marked decrease in ALT, TBR, CK, potassium, calcium and platelet count and increase in urea, creatinine, cholesterol, LDL-C, glucose and leucocytes count. However further studies on more animals and human beings are necessary to defend the utilization of multiple drugs.

These studies provides detailed evaluation of dug interaction and adverse effect of cumulative drug therapy; such observations are of undisputed importance but it should not be disregarded that pathway of drug metabolism in man may be quite dissimilar from that which has been determined in many species of laboratory animal, hence trial in man is the only valid way of establish drug interactions, before reaching to any final conclusion. However the risk of adverse drug reactions and drug interactions can be reduced by forming drug information centers, continuous medical education and incorporation of adverse drug reaction reporting into the clinical activities of the physicians (Oshikoya and Awobusuyi, 2009).

10. References

Adrogue, H.J. and Madias N.E. (2007). Sodium and potassium in the pathogenesis of hypertension. The New England Journal of Medicine, Vol. 356, No.19, pp.1966-1978

Astegiano, M., Sapone, N., Demarchi, B., Rossetti, S., Bonardi, R. and Rizzetto M. (2004). Laboratory evaluation of the patient with liver disease. European Review for Medical and Pharmacological Sciences, Vol. 8, No.1, pp. 3-9

Aziz, H., Ali, S.M., Frances, P, Khan, M.I. and Hasan K.Z. (1994). Epilepsy in Pakistan: a population-based epidemiologic study. Epilepsia, Vol. 35, No.5, pp.950-958

Aziz, H., Güvener, A., Akhtar, S.W. and Hasan K.Z. (1997). Comparative epidemiology of epilepsy in Pakistan and Turkey: population-based studies using identical protocols. Epilepsia, Vol. 38, No. 6, pp. 716-722

Bialer, M. and White H.S. (2010). Key factors in the discovery and development of new antiepileptic drugs. Nature Reviews Drug Discovery, Vol. 9, No. 1, pp. 68-82

Boden, W.E. and Taggart, D.P. (2009). Diabetes with coronary disease-a moving target amid evolving therapies? The New England Journal of Medicine, Vol. 360, No. 24, pp. 2570-2572

Brodie, M.J. and Dichter, M.A. (1996). Antiepileptic drugs. The New England Journal of Medicine, Vol. 334, pp. 168-175

Brown, M.S. and Goldstein, J.L. (1984). How LDL receptors influence cholesterol and atherosclerosis. Scientific American, Vol. 251, pp. 52-60

Campbell, M.F. and Allain, T.J. (2006). Amiodarone, sunlight avoidance and vitamin D deficiency. British Journal of Cardiology, Vol.13, No. 6, pp. 430-431

Chobanian, A.V. (2008). Does it matter how hypertension is controlled? The New England Journal of Medicine, Vol. 359, No. 23, pp.2485-2488

Cooper, M.S. and Gittoes, J.L. (2008). Clinical review: diagnosis and management of hypocalcaemia. British Medical Journal, Vol. 336, No. 7656, pp.1298-1302

Cutler, J.A., Sorlie, P.D., Wolz, M., Thom. T., Fields, L.E. and Roccella, E.J. (2008). Trends in hypertension prevalence, awareness, treatment, and control rates in United States adults between 1988–1994 and 1999–2004. Hypertension, Vol. 52, No.5, pp.818-827

Dichter, M.A. and Brodie, M.J. (1996). New antiepileptic drugs. The New England Journal of Medicine, Vol. 334, pp. 1583-1590

Dickson, M.E. and Sigmond, C.D. (2006).Genetic basis of hypertension. Hypertension, Vol.48, pp.14-20

Dietschy, J.M. (1998). Dietary fatty acids and the regulation of plasma low density lipoprotein cholesterol concentrations. Journal of Nutrition, Vol. 128, No. 2, pp.444S-448S

Dixon, N. (2002). Pharmacists as a part of an extended diabetes team. Pharmaceutical Journal, Vol. 268, No. 7192, pp. 469-470

Epstein, M. and Sowers, J.R. (1992). Diabetes mellitus and hypertension. Hypertension, Vol.19, No. 5, pp.403-418

Feroz, Z., Khan, R.A. and Afroz, S. (2009). Effect of multiple drug administration on gross toxicities and electrolytes. Pakistan Journal of Pharmacology, Vol. 26, No.2, pp. 33-39

Feroz, Z., Khan, R.A. and Afroz, S. (2011a). Adverse effects of anti-epileptic, anti-hypertensive, anti-diabetic and anti-arrhythmic drugs on hematological and hepatic parameters. Latin American Journal of Pharmacy, Vol. 30, No. 2, pp. 229-236

Feroz, Z., Khan, R.A. and Afroz, S. (2011b). Cumulative toxicities on lipid profile and glucose following administration of anti-epileptic, anti-hypertensive, anti-diabetic and anti-arrhythmic drugs. Pakistan Journal of Pharmaceutical Sciences, Vol. 24, No. 1, pp. 47-51

Feroz, Z., Khan, R.A., Mirza, T. and Afroz, S. (2010). Adverse effects of cumulative administration of anti-epileptic, anti-hypertensive, anti-diabetic and anti-arrhythmic drugs on renal and cardiac parameters. International Journal of Medicobiological Research, Vol. 1, No. 1, pp. 39-47

Fisher, R.S., van Emde Boas, W., Blume, W., Elger, C., Genton, P., Lee, P. and Engel, J. (2005). Epileptic seizures and epilepsy: definitions proposed by the international league against epilepsy (ILAE) and the international bureau for epilepsy (IBE). Epilepsia, Vol. 46, No. 4, pp. 470-472

French, J.A. and Pedley, T.A. (2008). Initial management of epilepsy. The New England Journal of Medicine, Vol.359, No. 2, pp.166-176

Giannini, E.G., Testa, R. and Savarino, V. (2005). Liver enzyme alteration: a guide for clinicians. Canadian Medical Association Journal, Vol. 172, No. 3, pp. 367-79

Goh, K.P. (2004). Management of hyponatremia. American Family Physician, Vol. 69, No. 10, pp. 2387-2394

Haffner, S.M., Lehto, S., Ronnemaa, T., Pyorala, K. and Laakso, M. (1998). Mortality from coronary heart disease in subjects with type 2 diabetes and in non-diabetic subjects with and without prior myocardial infarction. The New England Journal of Medicine, Vol. 339, pp.229-234

Hayat, A.S. and Shaikh, N. (2010). Barriers and myths to initiate insulin therapy for type 2 diabetes mellitus at primary health care centres of Hyderabad district. World Applied Sciences Journal, Vol. 8, No. 1, pp.66-72

Herlong (1994). Approach to the patient with abnormal liver enzymes. Hospital Practice, Vol. 29, No.11, pp. 32-38

Hoque, M.A., Islam, S., Khan, A.M., Aziz, R. and Ahasan H.N. (2009). Achievement of awareness in a diabetic population. Journal of Medicine. Vol. 10, No. 1, pp. 7-10

Howard, B.V., Robbins, D.C., Sievers, M.L., Lee, E.T., Rhoades, D., Devereux, R.B., Cowan, L.D., Gray, R.S., Welty, T.K., Go, O.T. and Howard, W.J. (2000). LDL cholesterol as a strong predictor of coronary heart disease in diabetic individuals with insulin resistance and low LDL. Arteriosclerosis Thrombosis and Vascular Biology, Vol. 20, PP.830-835

Hubscher, S.G. (2006). Histological assessment of the liver. Medicine, Vol. 35, No.1, pp. 17-21

Jawad, F. (2003). Diabetes in Pakistan. Diabetes Voice, Vol. 48, No. 2, pp. 12-14

Kassab, E., McFarlane, S.I. and Sowers, J.R. (2001).Vascular complications in diabetes and their prevention. Vascular medicine, Vol.6, No. 4, pp. 249-255.

Keen, H. (1998). Impact of new criteria for diabetes on pattern of disease. Lancet, Vol. 352, No. 9133, pp. 1000-1001

Khatri, I.A., Iannaccone, S.T., Ilyas, M.S., Abdullah, M. and Saleem, S. (2003) Epidemiology of epilepsy in Pakistan: review of literature. The Journal of Pakistan Medical Association, Vol. 53, No.12, pp. 594-596

Kim, Y.J. (2008). Interpretation of liver function tests. Korean Journal of Gastroentrology, Vol. 51, No. 4, pp. 219-224

Kratz, A., Lewandrowski, K.B., Siegel, A.J., Chun, K.Y., Flood, J.G., Cott, E.M.V. and Lee-Lewandrowski, E. (2002). Effect of marathon running on hematologic and biochemical laboratory parameters, including cardiac markers. American Journal of Clinical Pathology, Vol. 118, No. 6, pp. 856-863

Krier, B.P., Parker, R.D., Grayson, D. and Byrd, G. (1999). Effect of diabetes education on glucose control. Journal of the Louisiana State Medical Society, Vol. 151, No. 2, pp. 86-92

Kumar, U., Sharan, A. and Kamal, S. (2003). Raised serum lactate dehydrogenase associated with gangrenous small bowel volvulus: A case report. The Indian Journal of Clinical Biochemistry, Vol. 18, No. 2, pp. 6-7

Kwan, P. and Brodie, M.J. (2000). Early identification of refractory epilepsy. The New England Journal of Medicine, Vol. 342, No. 5, pp. 314-319

Lardinois, C.K. (1995). Nutritional factors and hypertension. Archives of Family Medicine, Vol. 4, No. 8, pp. 707-713

LaRoche, S.M. and Helmers, S.L. (2004). The new antiepileptic drugs: scientific review. The Journal of the American Medical Association, Vol. 291, No. 5, pp. 615-620

Lesko, L.J. (1989). Pharmacokinetic drug interaction with amiodarone. Clinical Pharmacokinetics, Vol. 17, No. 2, pp. 130-140

Lim, S.H. (2004). Epidemiology and etiology of seizures and epilepsy in the elderly in Asia. Neurology Asia, Vol. 9, No. 1, pp. 31-32

Mac, T.L., Tran, D.S., Queta, F., Odermatt, P., Preux, P.M. and Tan, C.T. (2007). Epidemiology, aetiology and clinical management of epilepsy in Asia. The Lancet Neurology, Vol. 6, pp. 533-543

Mahomed, I.M. and Ojewole, J.A. (2003). Hypoglycemic effect of Hypoxis hemerocallidea corm (African potato) aqueous extract in rats. Methods and Findings in Experimental and Clinical Pharmacology, Vol. 25, No. 8, pp. 617-623

Mant, M.J., Thomas, C., Philip, G. and Garner, K.E. (1979). Severe thrombocytopenia probably due to acute folic acid deficiency. Critical Care Medicine, Vol. 7, No. 7, pp. 297-300

McHugh, J.C. and Norman. D. (2008). Epidemiology and classification of epilepsy: gender comparisons. International Review of Neurobiology, Vol. 83, PP. 11-26

McMillan, R. (2007). Hemorrhagic disorders: Abnormalities of platelet and vascular function (L. Goldman & D. Ausiello, ed.), W.B. Saunders Company, Philadelphia, pp. 1289-1301

Mendez, M., Parera, V., Salamanca, R.E.D. and Batlle, A. (1999). Amiodarone is a pharmacologically safe drug for porphyrias. General Pharmacology, Vol. 32, No. 2, pp. 259-263

Muchmore, D.B., Springer, J. and Miller, M. (1994). Self monitoring of blood glucose in overweight type 2 diabetes patients. Acta Diabetologica, Vol. 31, No. 4, pp. 215-219

Nadkarni, S., Lajoie, J. and Devinsky, O. (2005). Current treatments of epilepsy. Neurology, Vol. 64, No.12 Suppl 3. pp. S2-S11

Naughton, C.A. (2008). Drug induced nephrotoxicity. American family physician, Vol. 78, No. 6, pp. 743-750

Nicolucci, A., Greenfield, S. and Mattke, S. (2006). Selecting indicators for the quality of diabetes care at the health systems level in OECD countries. International Journal for Quality in Health Care, Vol. 18, No. 1, pp. 26-30

Nolan, P.E., Erstad, B.L., Hoyer, G.L., Bliss, M., Gear, K. and Marcus, F.I. (1990). Steady state interaction between amiodarone and phenytoin in normal subjects. American Journal of Cardiology, Vol. 65, No. 18, pp. 1252-1257.

Oshikoya, K.A. and Awobusuyi, J.O. (2009). Perceptions of doctors to adverse drug reaction reporting in a teaching hospital in Lagos, Nigeria. BMC Clinical Pharmacology, Vol. 9, pp. 14.

Pelonero, A.L., Levenson, J.L. and Pandurangi, A.K. (1998). Neuroleptic malignant syndrome: a review. Psychiatric Services, Vol. 49, pp. 1163-1172

Peralta C.A., Shlipak, M.G., Wassel-Fyr, C., Bosworth, H., Hoffman, B., Martins, S., Oddone, E. and Goldstein, M.K. (2007). Association of antihypertensive therapy and diastolic hypotension in chronic kidney disease. Hypertension, Vol. 50, No. 3, pp. 474-480

Perazella, M. A. (2000). Drug-induced hyperkalemia: old culprits and new offenders. The American Journal of Medicine, Vol. 109, No. 4, pp. 307-314

Petkova, V., Ivanova, A. and Petrova, G. (2006). Education of patients with diabetes in the community pharmacies (pilot project in Bulgaria). Journal of Faculty of Pharmacy, Ankara, Vol. 35, No. 2, pp. 111-124

Pratt, D.S. and Kaplan, M.M. (2000). Evaluation of abnormal liver-enzyme results in asymptomatic patients. The New England Journal of Medicine, Vol. 342, No.17, pp. 1266-1271

Ramaiah, S.K. (2007). A toxicologist guide to the diagnostic interpretation of hepatic biochemical parameters. Food and Chemical Toxicology, Vol. 45, No. 9, pp. 1551-1557

Renner, E.L. and Dallenbach, A. (1992). Increased liver enzymes: what should be done? Therapeutische Umschau, Vol. 49, No. 5, pp. 281-286

Richer, M. and Robert, S. (1995). Fatal hepatotoxicity following oral administration of amiodarone. Annals of Pharmacotherapy, Vol. 29, No. 6, pp. 582-586

Riste, L., Khan, F. and Cruickshank, K. (2001). High Prevalence of type 2 diabetes in all ethnic groups, including Europeans, in a British inner city: relative poverty, history, inactivity or 21st century Europe? Diabetes Care, Vol. 24, No. 8, pp. 1377-1383

Ross, R. (1993). The pathogenesis of atherosclerosis: a perspective for the 1900s. Nature, Vol. 362, No. 6423, pp. 801-809.

Sakals, S., Peta, G.R., Fernandez, N.J. and Allen, A.L. (2006). Determining the cause of hypercalcemia in a dog. Canadian Veterinary Journal, Vol. 47, No. 8, pp. 819-821.

Schafer, A.I. (2004). Thrombocytosis. The New England Journal of Medicine, Vol. 350, pp. 1211-1219

Smith, A.F., Radford, D., Wong, C.P. and Oliver, M.F. (1976). Creatine kinase MB isoenzyme studies in diagnosis of myocardial infarction. British Heart Journal, Vol. 38, No. 3, pp. 225-232

Sowers, J.R. (2004). Treatment of hypertension in patients with diabetes. Archives of Internal Medicine, Vol. 164, No. 17, pp. 1850-1857

Sowers, J.R. and Epstein, M. (1995). Diabetes mellitus and associated hypertension, vascular disease and nephropathy. Hypertension, Vol. 26, No. 6 Pt 1, pp. 869-879

Sowers, J.R. and Zemel, M.B. (1990). Clinical implications of hypertension in the diabetic patient. American Journal of Hypertension, Vol. 3, No. 5 Pt 1, pp. 415-424

Strewler G.J. (2000). The physiology of parathyroid harmone related protein. The New England Journal of Medicine, Vol. 342, No. 3, pp. 177-185

Stults, B. and Jones, R.E. (2006). Management of hypertension in diabetes. Diabetes spectrum, Vol.19, N0.1, pp. 25-31.

Tenenbaum, A., Fisman, E.Z., Boyko, V., Goldbourt, U., Graff, E., Shemesh, J., Shotan, A., Reicher-Reiss, H., Behar, S. and Motro, M. (1999). Hypertension in diet versus pharmacologically treated diabetics. Hypertension, Vol. 33, No. 4, pp. 1002-1007

Thapa, B.R. and Walia, A. (2007). Liver function tests and their interpretation. The Indian Journal of Pediatrics, Vol. 74, No. 7, pp. 663-671

Usdin, Y.S., Sausville, E.A., Hutchins, J.B., Thomas, K. and Woosley, R.L. (1996). Amiodarone-induced lymphocyte toxicity and mitochondrial function. Journal of cardiovascular pharmacology, Vol.28, No. 1, pp. 94-100.

Varon, J. and Marik, P.E. (2003). Clinical review: The management of hypertensive crises. Critical care, Vol. 7, No.5, pp. 374–384.

Wasay, M. and Jabbar, A. (2009). Fight against chronic diseases (high blood pressure, stroke, diabetes and cancer) in Pakistan; cost-effective interventions. Journal of Pakistan Medical Association, Vol. 59, No. 4, pp. 196-197

Watanabe, M., Okamura, T., Kokubo, Y., Higashiyama, A. and Okayama, A. (2009). Elevated serum creatine kinase predicts first-ever myocardial infarction: a 12-year population-based cohort study in japan, the suita study. International Journal of Epidemiology, Vol. 38, No. 6, pp. 1571-1579

Weinberger, I., Rotenberg, Z., Fuchs, J., Ben-Sasson, E. and Agmon, J. (1987) Amiodarone induced thrombocytopenia. Archives of internal medicine, Vol. 147, No. 4, pp. 735-736

Wolbrette, D., Naccarelli, G., Curtis, A., Lehmann, M. and Kadish, A. (2002). Gender differences in arrhythmias. Clinical Cardiology, Vol. 25, No. 2, pp. 49-56

Yiu, K.H. and Tse, H.F. (2008).Hypertension and cardiac arrhythmias: a review of the epidemiology, pathophysiology and clinical implications. Journal of human hypertension, Vol. 22, No.6, pp. 380-388.

Zanella, MT., Kohlmann, O. and Ribeiro, A.B. (2001). Treatment of obesity hypertension and diabetes syndrome. Hypertension, Vol. 38, No. (3 Pt 2), pp. 705-708

Zimetbaum, P. (2007). Amiodarone for atrial fibrillation. The New England Journal of Medicine, Vol. 356, No. 9, pp. 935-941

Part 4

Biochemical Methods
for Food Preservation and Safety

7

Molecular Characterization and Serotyping of *Listeria monocytogenes* with a Focus on Food Safety and Disease Prevention

I.C. Morobe[1,2,4], C.L. Obi[1,3], M.A. Nyila[2], M.I. Matsheka[4] and B.A. Gashe[4]
[1]*Walter Sisulu University, Nelson Mandela Drive, Mthatha, Eastern Cape,*
[2]*School of Agriculture and Life Sciences, Department of Life and Consumer Sciences,*
University of South Africa, Pretoria,
[3]*Division of Academic Affairs and Research, Walter Sisulu University, Eastern Cape,*
[4]*Department of Biological Sciences, University of Botswana, Gaborone,*
[1,2,3]*South Africa*
[4]*Botswana*

1. Introduction

Listeria monocytogenes is a gram-positive, non-spore forming, facultative bacterium that is now an established food-borne pathogen known for causing the disease listeriosis in humans. Apart from displaying typical symptoms associated with gastrointestinal infections, listeriosis is also characterized by flu-like symptoms. The incidence of listeriosis is low in the general population, but what is significant is the very high fatality rate which can range from 20 to 30% in immune-compromised people (Mead *et al.*, 2006). In a report by Mead *et al.*, (2006) approximately 2500 cases of Listeriosis are recorded annually in the United States, resulting in 500 deaths. The primary mode of infection is through the ingestion of contaminated food. Therefore, food serves as an important medium in the transmission of infection and to date still plays a critical role in the propagation and perpetuation of cases of listeriosis the world over.

The ubiquitous nature of *L. monocytogenes* in the environment poses a challenge in reducing cases of listeriosis. The food industry is incapacitated in producing food free of this pathogen. Its wide distribution increases the chances of cross–contamination between appliances or several products during processing. *Listeria* also has the ability to colonize surfaces, forming biofilms that remain attached to equipment used in food production (Wong *et al.*, 1998). The organism ability to grow at low water activity, low pH (Buchanan *et al.*, 2000) as well as refrigerated vacuum packed food products (Duffy *et al.*, 1994) makes it difficult to eliminate from retail food products. *L. monocytogenes* is also problematic due to its resistance to multiple antibiotics, which makes it difficult to treat (Charpentier *et al.*, 1995).

Listeria Monocytogenes is divided into at least 12 serotypes (1/2a, 1/2b, 1/2c, 3a, 3b, 3c, 4a, 4b, 4c, 4d, 4e, and 7). The virulence of *Listeria monocytogenes* seems to be serotype dependent with serotypes 1/2a, 1/2c, 1/2b and 4b being involved in 98% of documented human listeriosis cases. The 4a and 4c serotypes of lineage are rarely associated with outbreaks of disease despite frequent isolation from a variety of food and environmental samples (Wiedmann *et al.*, 1997). Therefore, serotyping can provide useful information on the potential risk posed by *Listeria monocytogenes* isolates from various sources.

The conventional serotyping slide agglutination technique has been uses with great success in diagnostic and epidemiological investigations but is not routinely used because of the cost factor associated with the requirement of purchasing the whole spectrum of type specific antisera. Lack of standardization of reagents is known to hinder reproducibility of the technique. An enzyme–linked immunosorbent assay (ELISA) with the prospect of rapid serotyping of *L. monocytogenes* has been developed, but lack of concordance between the ELISA and agglutination results have been reported (Borucki *et al.*, 2003).This limitation has spurred a quest for alternative based typing techniques with universal application. Consequently *L. monocytogenes* PCR based serotyping methods have been described (Borucki *et al.*, 2003; Doumith *et al.*, 2004; Chen and Knabel, 2007) and chart a way into the establishment of a user friendly DNA based serotyping system. *L. monocytogenes* can now be further classified into three evolutionary lineages. Lineage I encompasses serotypes 1/2b, 3b, 4b, 4d and 4, lineage II includes serotypes 1/2a, 1/2c, 3a, and 3c; and lineage III comprises serotypes 4a, 4c as well as 4b.

While the PCR based serotyping methods play an important role in screening of subgroups of epidemiological importance, they need to be used in conjunction with sub-typing methods with higher discriminatory power in order to effectively track outbreak strains. Automated ribotyping of *Listeria monocytogenes* is an alternative DNA based typing technique of favor that has been used with great success in several investigations. However, this method has a prohibitive cost element, in that specialized equipment has to be purchased to analyze the results. PCR based typing methods remain an attractive cost effective option in sub-typing of bacterial isolates and have an added advantage of being rapid. REP-typing is a highly discriminatory typing technique has been used with success in epidemiologically surveillance of pathogenic bacteria of importance, including *Listeria* (Jersek *et al.*, 1998; Wojciech *et al.*, 2004).

Even though a past study by Manani *et al.*, (2006) reported the occurrence of *Listeria monocytogenes* in frozen vegetables in this country, there is little data on the occurrence of this pathogen in foods available to consumers in Botswana. The REP-typing method described by Jersek *et al*, (1998) together with the modified PCR serotyping method initially described by Borucki *et al.*, (2003), were used to determine the of presence of Listeria serotypes of clinical significance in various retail food products in Botswana. These methods have predominantly been used in isolates from Europe and North America (Chen and Knabel, 2007) and this study provided the opportunity to assess the robustness and appropriateness of the genetic markers underpinning these DNA based typing techniques in characterizing isolates from a geographically distinct region.

2. Literature review

2.1 Listeria disease

Listeria includes 6 different species, *L. monocytogenes*, *L. ivanovii* sub-species, *ivanovii*, *L. innocua*, *L. welshimeri*, *L. seegligeri* and *L. grayi*. Both *L. monocytogenes L. innocua* and *ivanovii*

L. are pathogenic, but only *L. monocytogenes* is associated with humans and animal illness (Rodriguez-Lazaro *et al.,* 2004). All these species are psychotrophic and widely spread in the environment. *L. innocua* is a major contaminant of vegetable surfaces and equipment or machinery (Aguado *et al.,* 2004).

Listeria monocytogenes appears to be a normal resident of the intestinal tract in humans. Studies carried out by examination of faecal samples, have found out that approximately 5 to 10 % of the general population are carriers of *L. monocytogenes.* This observation is thought to partially explain why antibodies to Listeria species are common in healthy individuals. Thus, because of the high rate of clinically healthy carriers, the presence of *L. monocytogenes* in faeces is not necessarily an indication of an infection (Farber *et al.,* 1991). Listeriosis is clinically defined when the organism is isolated from blood, cerebrospinal fluid and even placenta and foetus in cases of abortion. In a study of the duration of faecal excretion, shedding patterns were found to be erratic among different individuals with some carriers found to shed the organism for long periods. Among animals the carrier rate is generally considered to be 1 to 5 %, although recent studies involving newer methods for isolating *Listeria* species have indicated much higher carriage rates (Farber *et al.,* 1991).

Ingestion of food contaminated with *L. monocytogenes* can result in symptoms characteristic of listeriosis. Studies have shown that the number of *Listeria monocytogenes* cells can rise following refrigeration from fewer than 100 cells per gram, and this is the dose that is generally accepted for healthy people (Huss *et al.,* 2000; Buchanan *et al.,* 2000). Individuals who are particularly susceptible to this condition are immune-compromised individual (as in HIV/AIDS infection), pregnant women, persons with low stomach acidity, newborn babies, cancer patients, alcoholics, drug abusers, patients with corticosteroid therapy and the elderly (McLauchlin *et al.,* 2004; Rodriguez – Lazaro et *al.,* 2004). Most healthy individuals experience flu-like symptoms. The manifestation of listeriosis includes septicemia, meningitis (meningoencephalitis), encephalitis and intrauterine or cervical infections in pregnant women. According to Mead and his colleagues (1999), Infection acquired in early pregnancy may lead to abortion, still birth or premature delivery. When listeriosis is acquired late in pregnancy it can be transmitted transplacentally and lead to neonatal listeriosis. The onset of the aforementioned disorders is usually preceded by influenza-like symptoms including persistent fever followed by nausea, vomiting and diarrhea, particularly in patients who use antacid or cimetidine. The onset time to serious forms of listeriosis ranges from a few days to 3 weeks, while the onset time to gastrointestinal symptoms is greater than 12 hours (>12hours).The severe form of the disease has a high fatality rate of 30 %.

2.2 Determinants of *Listeria* virulence

Listeria monocytogenes produces an exotoxin listeriolysin (LLO) which is major virulence factor and its secretion is essential for promoting the intracellular growth and T- cell recognition of the organism (Farber *et al.,* 1991). The hemolysin of *L. monocytogenes* is recognized as a key agent in human neutrophil activation. The stimulation of these phagocytes, however, requires additional listerial virulence factors of which PIcA may play a prominent role (Karunasagar *et al.,* 1993). The presence of the listeriolysin gene is restricted to the species *L. monocytogenes.* Beside the characterized Listeriolysin encoded by the hly gene, *L. monocytogenes* also produces two other hemolysins; phosphatidylinositol-specific phospholipase C (Pl-PLC) and phosphatidylcholine-specific phospholipase C (PC-PLC). Unlike the LLO which lyses host cells by pore formation, these virulence factors act by disrupting the membrane lipids. The bacterium also produces zinc (2+) dependent protease, which acts like an exotoxin.

There are six *Listeria monocytogenes* virulent genes, namely; *prfA*, *pclA*, *hlyA*, *mpl*, *actA*, and *plcB* located together in one virulence gene cluster between the house keeping gene *idh* and *prs*. The *actA* gene product is a surface protein required for intercellular movement and cell to cell spread through bacterially induced actin polymerization. The virulence of *L. monocytogenes* is multi-factorial. Other factors affecting the pathogenicity of *L. monocytogenes* are, iron compounds, catalase and superoxide dismutase, and surface components. The virulence of the organism may be affected by its growth temperature. Growth of *L .monocytogenes* at a reduced temperature (4°C) increases its virulence intravenously. This phenomenon may affect the virulence of the organism in refrigerated foods.

2.3 Treatment

When infection occurs during pregnancy, antibiotics given promptly to the pregnant women can often prevent infection of the fetus or new born. In general, isolates of *L. monocytogenes*, as well as strains of other Listeria species, are susceptible to a wide range of antibiotics except tetracycline, erythromycin, streptomycin, cephalosporins, and fosfomycin (Charpentier *et al.*, 1999). The treatment of choice for listeriosis remains the administration of ampicillin, penicillin G combined with an aminoglycoside and gentamycin. The association of trimethoprim with sulphonamide, such as sulfamethaxazole in co–trimoxazole, is a second choice therapy (Charpentier *et al.*, 1999). The most active agent in the combination is trimethoprim, which is synergized by sulfamethaxazole. Most isolates from clinical as well foodborne and environmental sources are susceptible to the antibiotics active against gram positive bacteria (Yucel *et al.*, 2005).

2.4 Epidemiology and occurrence of *Listeria*

The incidence of listeriosis appears to be on the increase worldwide with a significant number of cases, especially in Europe. The annual endemic disease rate varies from 2 to 15 cases per million populations, with published rates varying from 1.6 to a high rate of 14.7 in France for 1986. *Listeria* has been isolated sporadically from wide variety of sources and listeriosis outbreaks that have occurred in the past have highlighted contaminated food as the main source of transmission. A wide range of foods such as salads, seafood's, meat, and dairy have been implicated in listeriosis (Huss *et al.*, 2000). Usually the presence of *Listeria* species in food is thought to be an indicator of poor hygiene (Manani *et al.*, 2006). A variety of ready- to–eat food products, such as frozen or raw vegetables, milk and milk products, meat and meat products and seafood support the growth of *Listeria monocytogenes*. These foods are considered of high risk due to the ability of listeria to grow and survive in them (Kunene *et al.*, 1999). However, there are other products, traditionally considered of low risk, which have recently been linked to listeriosis transmission, such as the large listeriosis outbreak reported in Italy due to the consumption of corn. Though no fatalities occurred, more than 1500 people were affected (Aguado *et al.*, 2004). There is no doubt that the susceptible population is increasing, as there is a steady increase in numbers and types of foods in which *L. monocytogenes* is isolated.

In Africa the incidence of listeriosis have also been reported in countries like Zambia where 85 cases of meningitis due to *Listeria* were reported (Chintu *et al.*, 1975). In Togo, 8 out of 342 healthy slaughter animals were positive for *L. monocytogenes* (serovars 1/2a and 4b) isolated from the intestinal lymph nodes (Hohne *et al.*, 1975). In Northern Nigeria 27% mortality due to *L. monocytogenes* (serovar; 4) was reported (Onyemelukwe *et al.*, 1983). Listeria organisms are documented to be zoonotic with one of the sources of infection being the domestic fowl. *L. monocytogenes* can be found on poultry carcasses and in poultry processing plants. The prevalence of pathogens in chickens in many countries is well documented but their presence in South African poultry products has not been extensively reported on. Two studies investigating contamination of food available from street vendors in Johannesburg have been documented (Mosupye and Von Holly 1999; 2000). The carriage of *Listeria monocytogenes* and other *Listeria* in indigenous birds has not been documented in Kenya (Njagi *et al.*, 2004).

2.5 Antimicrobial resistance in *L. monocytogenes*

Listeria monocytogenes, as well as other *Listeria* spp., are usually susceptible to a wide range of antibiotics (Charpentier *et al.*, 1995). However, evolution of bacterial resistance towards antibiotics has been accelerated considerably by the selective pressure exerted by over-prescription of drugs in clinical settings and their heavy use as promoters in animals husbandry (Charpentier *et al.*, 1995).Therefore, it was not unexpected when the first multi-resistant strain of *L. monocytogenes* was isolated in France in 1988 (Poyart-Salmeron *et al.*, 1990) and since then multi-resistant *L monocytogenes* strains have been recovered from food, the environment and sporadic cases of human listeriosis (Charpentier *et al.*, 1995). Antibiotics to which some *L. monocytogenes* strains are resistant to include tetracycline, gentamicin, penicillin, ampicillin, streptomycin, erythromycin, kanamycin, sulfonamide, trimethoprim, and rifampicin (Charpentier and Courvalin, 1999).

Tetracycline resistance has been the most frequently observed resistance phenotype among *L. monocytogenes* isolates (Charpentier *et al.*, 1995; Charpentier and Courvalin, 1999). Six classes of tetracycline-resistance genes; *tet*(K), *tet*(L), *tet*(M), *tet*(O), *tet*(P), and *tet*(S) have been described in Gram positive bacteria (Charpentier et al., 1995). However, only *tet*(L) and *tet*(S) have been identified in *L. monocytogenes* (Poyart-Salmeron *et al.*, 1992; Charpentier and Courvalin, 1999). *Tet*(M) and *tet*(S) confer resistance by ribosomal protection, whereas the *tet*(L) gene codes for a protein which promotes active efflux of tetracycline from the bacteria. Transfer of resistance between *L. monocytogenes* can occur in the gastrointestinal tract of domestic animals where both species live and where sub-inhibitory levels of tetracycline may be expected. In fact, tetracyclines are second most commonly used antibiotics worldwide. They are used extensively in animal foodstuffs, especially for poultry, and it is noteworthy that tetracycline resistance was the single most common resistance marker in food-borne *L. monocytogenes* isolated from chicken and turkey (Chopra *et al.*, 2001). Antibiotic resistance in *L. monocytogenes* is reaching an era where virtually all antibiotics will be rendered ineffective because of various mechanisms employed by *L monocytogenes* to counteract the therapeutic agents.

2.6 Isolation and detection of *L. monocytogenes*

The Genus *Listeria* includes 6 different species but only *L. monocytogenes* and *L. ivanovii* are known to be pathogenic. However, only *L .monocytogenes* is associated with humans and animal illness (Rodriguez-Lazaro *et al.*, 2004).In the past strains were classified to species level using morphological characteristics and biochemical tests (suspect colonies, motility, catalase, hemolysins, CAMP and API *Listeria* identification system). Significant efforts have been dedicated to the development of enrichment media and protocols for *L. monocytogenes* isolation. Ideal enrichment media would facilitate recovery of injured *Listeria* cells and enrichment of *Listeria* species (*L. monocytogenes*) over competing microflora. In traditional culture–based assays, it becomes very difficult to detect *L. monocytogenes* at any level when it is greatly outnumbered by other *Listeria* species, such as *L. innocua*, which is in most cases present together with *L. monocytogenes* (Bille *et al.*, 1992). Specific–specific identification with biochemical standard methods, which include sugar fermentation or the CAMP test, is laborious and time consuming and can require 1 to 2 weeks for species identification. Moreover differentiation between species and strains is not always reached (Aguado *et al.*, 2003). A diagnostic scheme for the same day identification of food borne cells of *L. monocytogenes* has been proposed. Large representative colonies that emerge in 40 hours at 30°C are used as heavy inoculum on agar plates for the rapid determination of hemolytic activity and acidification of rhamnose and xylose. Additional tests consisting of cell phase-contrast microscopy, motility testing, the catalase production test and the KOH viscosity test in place of Gram staining have been employed in the rapid identification of *L. monocytogenes* (Borucki *et al.*, 2003).

The rapid identification of *L. monocytogenes* is important so that the appropriate antibiotic therapy can be initiated. Currently molecular methods that enable the identification of *Listeria* to the species level include; Random Amplified Polymorphic DNA Polymorphism, to discriminate *Listeria monocytogenes* from *Listeria innocua* (in the 16S rRNA genes), Polymerase Chain Reaction (PCR), real time PCR, Ligase Chain Reaction (LCR) for the detection of *Listeria monocytogenes* and Pulse-field fingerprinting of *listeria*, for detection of genomic divisions for *L. monocytogenes* and their correlation with serovars, and restriction endonuclease analysis (REA), have been employed to directly characterize the microorganism without the need for isolation (Wouters *et al.*, 1999). These new molecular methods may also improve the ability to diagnose pregnancy –associated disease and permit the rapid detection and control of *L. monocytogenes* in the food supply (Wiedmann *et al.*, 1993).

The Listeriolysin genes have also been used identification. DNA hybridization studies have shown that listeriolysin genes are found in *Listeria* species, such as *L. monocytogenes, L. ivanovii, L. seeligeri*. In the analysis of genomic DNA of Listeria by southern hybridization with *hlyA* probes all strains were isolated and digested with the restriction endonuclease Hind*III*. The 0.8-kb Bam*HI* probe that was made up entirely of sequences upstream of the listeriolysin gene was found to hybridize to *L. monocytogenes* strains irrespective of serotype, as well as to the *L. seeligeri* and *L. ivanovii* strains (Borucki *et al.*, 2003). Other methods that can be employed to detect listerolysin are; hemolysin assays and polyacrylamide gel electrophoresis, imuno-magnetic beads for listeria and *Listeria* exotoxin detection kits (Borucki *et al.*, 2003). Immunoblotting performed with affinity–purified antibody to literiolysin allowed the detection of this protein in supernatants of all three species. In this immunological assay two recombinants, (pLM47 and pLM48) were found to produce a polypeptide of 60KDa which cross-reacted with the antisera to produce a hemolytic phenotype on blood agar plates (Leimeister- Wachter *et al.*, 1992).

2.7 Typing of L. monocytogenes.

L. monocytogenes strains are serotyped according to variation in the somatic (O) and flagellar (H) antigens (Seeliger and Hohne, 1979). Thirteen L. monocytogenes serotypes (serovars) have been characterized in this species by using specific and standardized sera (Seeliger & Langer, 1979). Serovar identification by serological tests has remained popular. However, numerous molecular biology methods such as multiplex PCR (Borucki et al., 2003; Doumith et al., 2004; Chen and Knabel) have come to the fore in the characterization of L. monocytogenes serotypes. Three genetic lineages (I, II, III) for L. monocytogenes have been identified using these various molecular subtyping techniques. Epidemic clone I contains serotypes 1/2b, 3b, 3c, and 4b, lineage II contained serotypes 1/2a, 1/2c, and 3a and lineage III contained serotypes 4a and 4c (Chen and Knabel, 2007). Although most clinical isolates belong to serovars 1/2a, 1/2b, and 4b, the majority of strains which have caused large outbreaks are serovars 1/2a and 1/2b (Kathariou, 2000); Jacquet et al., 2002; Zhang and Knabel, 2005). Interestingly, although serotype 1/2a is the most frequently isolated from food, serotype 4b causes the majority of human epidemics (Zhang and Nabel., 2005). This suggests that serovar 4b may pose unique virulence properties. However, geographical differences in the global distribution of serotypes apparently exist.

Phage typing has proven to be a valuable epidemiological tool in investigations of many infectious diseases. Since the initial discovery phages specific for listeria species in 1945, several groups have assessed the usefulness of phage typing L. monocytogenes. Phages derived from both environment sources and lysogenic strains have been found. In isozyme typing, bacteria are differentiated by the variation in the electrophoretic mobility of any of a large number of metabolic enzymes. This technique is useful in either confirming or eliminating a common source as the cause of an outbreak of food-borne listeriosis (Farber et al., 1991).

DNA fingerprinting using restriction enzyme analysis (REA) has recently been used to characterize strains of L. monocytogenes causing outbreaks of listeriosis associated with Mexican–style soft cheese in Los Angeles, as well as the Nova Scotia and Switzerland outbreaks (Aguado et al., 2004). Plasmid typing has also been used in conjunction with DNA fingerprint to confirm a case of cross–infection with L. monocytogenes. However, this technique is of less importance since L .monocytogenes does not appear to carry plasmid. On the other hand L. innocua carry plasmids ranging in size from 3 to 55 MDa (Aguado et al., 2004). Monocine typing has recently been evaluated as a typing tool for L. monocytogenes. Although this technique is potentially promising as an epidemiological tool, only 59 and 56 % of serovars 1/2a and 4b were found to be producers of monocines. In one instance a pair of L. monocytogenes strains isolated from a mother and a newborn, which could not be phage typed, but proved to be identical by monocine typing (Baloga et al., 2006).

3. Materials and methods

3.1 Sampling

Samples were obtained randomly from selected supermarkets and street vendors in 5 Geographical areas of Gaborone Samples collected were; raw vegetables (cabbage) and salads, raw milk, cheese and meat (biltong). In this study 250 -300 samples per product were obtained. Samples were put in separate properly labeled sterile specimen bags and put into a cooler box containing ice packs. Gloves were worn to avoid cross- contamination between samples from different supermarkets and street vendors. Samples were transferred into properly labeled stomacher bags and then homogenized with the Stomacher (Seward 400, Tekmar, and Cincinnati Ohio, USA) set at medium speed. *Listeria monocytogenes* positive control (ATCC 19115) was purchased from South African Bureau of Standards (SABS).

3.2 Enrichment, culturing, morphological observation and biochemical identification

The homogenized samples were enriched by putting 25g of the sample into 225ml enrichment broth (Mast Diagnostics DM257) and incubated at 30°C for 48 hours on Innova 4000 Newbrunswick Scientific incubator shaker. A loop full of culture was subcultured after 48 hours onto Listeria Selective Agar (Oxoid M009, Basingstoke, UK) plates and then incubated at 37°C for 24 hours. Modified Listeria Selective Enrichment Supplement (SR206E, Oxoid, Basingstoke, and Hampshire, England) was added to Listeria Agar Base and dark brown colonies with black zones were subcultured on nutrient agar (Oxoid CM001) plates. A Gram stain was done on suspected colonies from a culture medium (Nutrient Agar). Gram positive short rods colonies picked were subcultured onto Tryptic Soya Agar (Merck, Darmstadt, Germany) slants. After 24 hour incubation, the slants were kept at 4°C. Other broth cultures were stored in 80% Tryptic Soya Broth and 20% Glycerol (i.e., 750µl Tryptose broth plus 250µl of 20% Glycerol) were put into a 2ml vial and kept at -82°C for subsequent steps.

3.2.1 Biochemical testing

A catalase test was performed to separate the listeria (catalase positive) from listeria (catalase negative). Picking a colony with sterile loop on a slide containing 3% hydrogen peroxide does this. The evolution of gas bubbles (oxygen) indicates a positive test. Sero-agglutination was carried out by a slide agglutination technique using commercially prepared *Listeria* spp. antisera (Oxoid Listeria test kit DR1126A, Basingstoke, New Hampshire, England). Agglutination patterns were linked to *Listeria* spp. following manufacturer's instructions. Isolates that were positive in the serology test were subjected to the API Listeria test (BioMerieux, Paris, France). This is the confirmatory test for the organism and it differentiates *L. monocytogenes* from other *Listeria* species.

3.3 Serotype identification by PCR

3.3.1 DNA extraction

L. monocytogenes strains were stored long term in tryptic soy broth (Merck, Darmstadt, Germany) with 20% glycerol at -82°C. Strains were then recovered by inoculating into tryptose soy broth and were grown overnight at 37 °C. Cells were harvested on a bench top centrifuge and genomic DNA extracted using Guanidium Thiocyanate chromosomal technique (Pitcher et al., 1989). 500µl of guanidium thiocyanate solution (60g Guanidium thiocyanate, 20ml 0.5M EDTA pH8,20ml deionized water and 5ml of 10% (w/v) N-Lauryl-Sarcosine Sodium salt, made up to 100ml with deionized water) was added and briefly mixed to lyse the cells. 250µl ice-cold 7.5M ammonium acetate was added, mixed and left on ice for 10 minutes. 500µl Chloroform/ Isopropanol (24:1) was added, mixed and span at 1200g for 10 minutes. 600µl supernatant was transferred to clean Eppendorf tubes and 400µl ice –cold isopropanol was added, mixed by gentle inversion of the Eppendorf tube and span at 6500g for 20 seconds to collect DNA. The pellet was washed three times with 200μl of ice-cold 70% alcohol, span at 6500g for 2 minutes to remove excess Guanidium thiocyanate. The pellet was then air dried at 4 °C for 7 minutes and resuspened in 30μl of TE buffer and kept at 4°C for subsequent steps.

3.3.2 PCR amplification

Amplification of serotype specific *hly* gene product of 214 bp and the Serogroup identification by multiplex PCR using primer pairs D1 and D2. PCR using reverse and forward primers (D1-Forwad; 5′ CGATATTTTATCTACTTTGTC 3′; D1-Reverse; 5′ TTGCTCCAAAGCAGGGCAT 3′ and D2-Forward; 5′ GCGGAGAAAGCTATCGCA 3′; D2-Reverse; 5′ TTGTTCAAACATAGGGCTA 3′) as described by Borucki and Call (2003). Reaction mixtures was made up to 50μl using the high pure PCR template kit (Fermentas) and Roche PCR core kit reagent according to manufacturer's instructions. Each reaction consisted of 50pmol of each primer and 50ng of DNA template with 2.5 units of Tag polymerase.

Amplification was carried out using Applied Biosystems GeneAmp 2400 thermocycler. PCR cycling conditions were as follows; 95°C for 3 minutes followed by 25 cycles (with D1 and D2 primers) 72°C for 1 minute followed by a final step of 72°C for 10 minutes after cycling was completed. The product size was resolved using electrophoresis through 1.8% agarose gels containing ethidium bromide and visualized on a UV transilluminator. During this experiment laboratory control strain of *L. monocytogenes* were used as a positive control, included in each group of samples undergoing analysis.

The strains that tested positive with D1 primers were further subjected to PCR using GLT primers (GLT-Forward 5′- AAA GTG AGT TCT TAC GAG ATT T-3′ and GLT-Reverse 5′-AAT TAG GAA ATC GAC CTT CT-3′). The PCR reaction conditions were as mentioned above but with a different PCR cycling protocol. Initial denaturation was carried out at 95°C for 5 minutes followed by 25 cycles of 45°C for 30 seconds and 72°C for 1 minute, followed by a final step of 72°C for 10 minutes after cycling was completed. PCR products were determined using electrophoresis through 1.8% agarose gel containing ethedium bromide and visualized on a UV transilluminator.

3.3.3 Lineage group classification by MAMA – PCR

MAMA primers were used to test strains that tested negative with GLT primers. The high pure PCR template kit (Fermentas) was used according to the manufacturer's instructions. Reaction mixtures contained primers (LM4-Forward (5′- CAG TTG CAA GCG CTTGGAGT-3′) and LMB-Reverse (5′- GTA AGT CTC CGA GGT TGC AA-3′) at a concentration of 50pmoles. MAMA-PCR amplification conditions were as follows; 10 minutes initial denaturation step, followed by 40 cycles of 0.5 minutes at 95°C, 1 minute at 55°C and 1 minute at 72°C, with a final extension step for 10 minutes at 72°C. Amplification product was electrophoresed on a 1.5% agarose gel containing 0.4μg/ml ethidium bromide at 60 volts for 90 minutes and visualized on a UV transilluminator.

The strains that tested positive with MAMA primers were subjected to PCR using ORF2110 primers   (Forward: 5′   -   AGTGGACAATTGATTGGTGAA-3′   and   Reverse: 5′-

CATCCATCCCTTACTTTGGAC-3') as described by Doumith *et al.* (2004) at a concentration of 50pmoles.ORF2110 - PCR amplification conditions were as follows; initial denaturation step at 94°C for 3 minutes followed by 35 cycles of 94°C for 0.40 minutes, 53°C for 1 minute and 72°C for 1 minute and one final cycle of 72°C for 7 minutes. Amplification product was electrophoresed on a 1.5% agarose gel containing 0.4µg/ml ethidium bromide at 60 volts for 90 minutes and visualized on a UV transilluminator.

3.4 Typing by repetitive element sequence – based PCR

Amplification of REP-PCR products was done using REP IR – I 5'– IIIICGICGICATCIGGC-3' and REP 2-1 5'-ICGICTTATCIGGCCTAC-3' primer pairs as described by Jersek et al., (1999). Reaction mixtures made up of 50µl using the high pure PCR template kit (Fermentas) and Roche PCR core kit reagents were used according to the manufacturer's instructions. Each reaction consisted of 50pmol of each primer and 50ng of DNA template with 2.5 units of Tag polymerase. REP – PCR cycling conditions were as follows; An initial denaturation at 95°C for 3 minutes followed by 30 cycles of 90°C for 30 seconds at 40°C for 1 minute , at 72°C for 1 minute and final cycle at 72°C for 8 minutes. The REP-PCR gene products were resolved into finger printing patterns on a 1.5% agarose gel (Roche) at 60Volts for 1 hour. Gel images captured on the Syngene Gene Genius BioImaging System (Cambridge, UK). Fingerprint patterns were considered different if there was a presence or absence of a band at a particular molecular weight. Variations in the brightness of the band, was not considered to constitute a difference. The gel images were subjected to cluster analysis using GelCompar software (Applied Maths, Kortrijk, Belgium). The similarity was performed using the Dice coefficient. A band matching tolerance of 1.0% was chosen.

4. Results

4.1 Isolation and identification of *L. monocytogenes*

From the five various food products (Cheese, raw milk, Biltong, frozen cabbage and Coleslaw salad), 57 isolates of *L. monocytogenes* were recovered from all food types except biltong. These isolates were identified as *Listeria* by using the seroagglutination test and positively identifies as *L. monocytogenes* using phenotypic and biochemical testing.

4.2 Serogroup identification by Polymerase Chain Reaction (PCR)

Serogroup identification by PCR was performed on all the 57 confirmed *L. monocytogenes* isolates. Using primer pairs D1 a PCR product of 214bp was for the entire strains analyzed. The PCR product of this size suggests serotypes belonging to phylogenetic lineage of division I and III, which comprise serogroups 4a, 1/2b, 3b, 4b, 4c, 4d and 4e. No PCR product was obtained for all 57 isolates using The D2 Lineage II specific primers.

To differentiate 1/2b and 3b serotypes from the rest of the members in division I, the strains that tested positive with D1 primers were subjected to PCR using GLT primers. Only 6 (10.52%) strains gave the expected PCR product size of 483b. In one specific case the primers gave an amplicon bigger than the expected PCR product (Fig 2 lane 4).

Strains that did not give any product (amplicon) or the expected PCR product size with GLT primers were assumed to belong to either division I or division III. To differentiate the

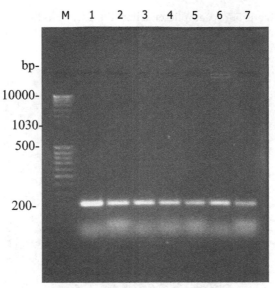

Fig. 1. Genomic DNA from *Listeria monocytogenes* strains subjected to multiplex PCR with primer pairs for D1 and D2. Showing the 214bp PCR product (lanes 1-7). Lane M, MassRuler ™ SM0403 (Fermentas).

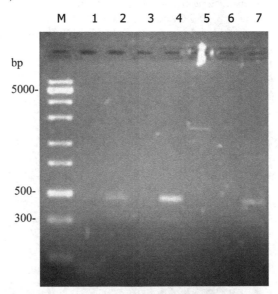

Fig. 2. Genomic DNA from *Listeria monocytogenes* strains subjected to PCR with GLT primers. Lane 2, 3 and 7 (ATCC 19115 strain) depicting isolates with the expected 483bp PCR product. Lane M, ZipRuler ™ SM1378 (Fermentas).

isolates belonging to serogroup 4b, 4d and 4e, the isolates which did not give a PCR product with GLT primers were again subjected to PCR with primers specific to ORF2110. Seventeen of the 57 isolates gave a 597bp amplicon when the PCR product was resolve on a 1.5% agarose gel. In certain instances unspecific priming was evident (Fig 3 lanes 4, 5 and 6).

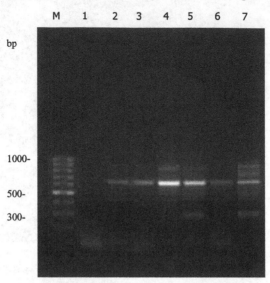

Fig. 3. Genomic DNA from *Listeria* monocytogenes strains subjected to PCR with primers specific to ORF 2110. Lane 2, 3, 4, 5 6 and 7(ATCC19115 strain) depicting isolates with the expected 596bp PCR product. Lane M, GeneRuler ™ SM1148 (Fermentas).

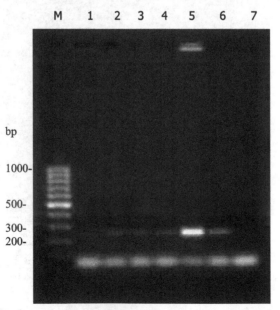

Fig. 4. Genomic DNA from *Listeria monocytogenes* strains subjected to PCR with MAMA-C primers. Lanes 2 to 6 show the expected 268bp PCR product. Lane M, GeneRuler™ SM 1148.

Furthermore the isolates that were GLT negative were subjected to PCR using MAMA-C primers to identify isolates belonging to division III. From the 51 isolates that were negative for GLT primers, 45 isolates gave a 268bp product with MAMA-C specific primers, indicating suggesting that they belonged to Division III. Included in the 45 positive isolates that were positive for PCR with MAMA-primers were the 17 isolates that had proved of positive for ORF2110 specific primers. The rhamnose fermentation test was positive for all 45 isolates suggesting that they belonged to division III.

Serogroup identification using PCR found that most isolates (49%) belonged to serogroups 4a and 4c. These isolates were found mostly in cabbage and salads. This was followed by isolates in serogroups 4b, 4d and 4e which comprised 30% of the isolates. Isolates within this group appeared in all food types except salads and cheese. Isolates with serogroups 1/2b and 3b were rare with four isolates appearing in salads and one isolate being picked up in cheese and milk respectively. Isolates belonging to division II were not detected at all because no isolates were positive for PCR using D2 specific primers. Four isolates in salads (S10, S31, S200, S258) and two (V208, V225) in cabbage were found to belong to division I, but could not be characterized further into serogroups because they proved to be negative for PCR serogroup identification using GLT and ORF2110 specific primers.

| Food type | Strains | Serotyping by PCR | | | | Serogroup |
		D1	GLT	ORF2110	MAMA-C	
Cheese	C6	+	+	-	-	1/2b and 3b
Milk	RM16	+	+	-	-	1/2b and 3b
	RM111,RM117	+	-	+	+	4b, 4d and 4e
Cabbage	V1,V61,V62,V63,V92,V106,V113 V121,V129,V131,V148,V157,V189 V243,V259,V262	+	-	-	+	4a and 4c
	V4,V5,V43,V78,V97,V166,V231 V238,V250	+	-	+	+	4b, 4d and 4e
	V208,V225	+	-	-	-	ND
Salad	S7,S43,S44,S55,S77,S81,S88,S90 S104,S125,S169,S176	+	-	-	+	4a and 4c
	S1,S24,S199,S221	+	+	-	-	1/2b and 3b
	S10,S31,S200,S258	+	-	-	-	ND

+ (positive), - (negative), ND (Not Determined)

Table 1. Serotyping of *L. monocytogenes* by PCR.

4.3 Typing by repetitive element sequence based PCR

From the 57 *L. monocytogenes* strains that were isolated, 41 were selected and typed using REP-PCR. DNA fingerprints obtained for all isolates had a maximum of five bands ranging from 200bp to 300bp (See figure 5. A, B, C and D). The DNA fingerprints were analyzed using a cluster analysis computer program to derive a dendrogram. A total of 5 clusters could be clearly distinguished at similarity level of more than 40% (Fig. 6) which are designated I, II, III, IV and V. The REP-PCR profiles of the isolates seemed to cluster according to food types, with most of the isolates from salad and cabbage falling into group I and V respectively. The three *L. monocytogenes* isolated from milk fell into group II while all five cheese isolates selected for REP-PCR analysis fell into group III. Group IV was heterogenous group with a balanced mixture of isolates from salad and cabbages. Two pairs of isolates from the two food types had identical REP-PCR profiles (S7 was similar to V131, while S10 was similar to V157). Cabbage had the most diverse REP-PCR profile types that appeared in 4 of the 5 REP-PCR profile groups. Only REP-PCR profile group I had isolates from one commodity (salads). There was no correlation between the REP-PCR clusters and serogroup or the three *L. monocytogenes* genetic lineages.

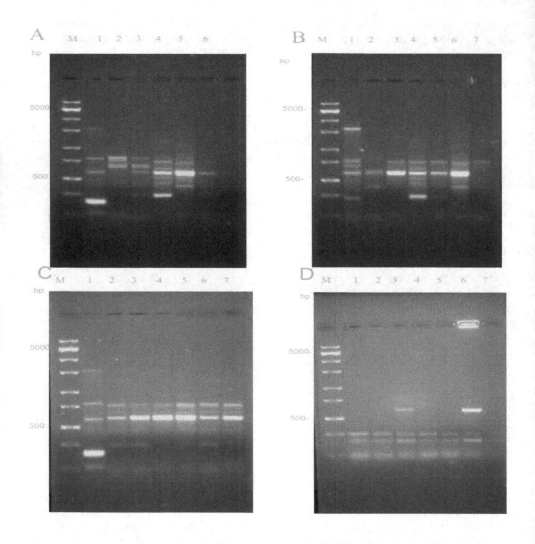

Fig. 5. REP – PCR fingerprints of *Listeria monocytogenes* isolates (lane 1-7). Lane M, ZipRuler ™ SM1378 (Fermentas). A and B shows the diverse profiles of isolates obtained from cabbage and salads. C and D shows isolates with similar REP-PCR profiles obtained from cabbage and salads respectively.

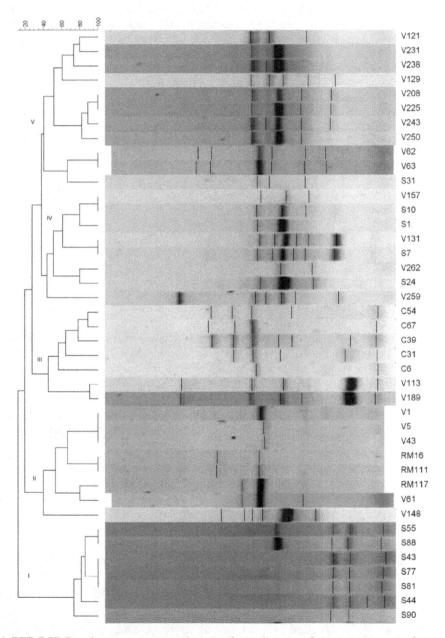

Fig. 6. REP-PCR Dendrogram representing similarity between *L. monocytogenes* isolates.

5. Discussion

In this study serogrouping by PCR suggests serotypes belonging to phylogenetic lineage of division I and III, which comprise serogroups 1/2b, 3b, 4a, 4b, 4c, 4d and 4e. Isolates within this group appeared in all food types except salads. It was found that isolates with 1/2b, and 3b were rare with only four isolates appearing in salad and one isolate being picked up in cheese and milk respectively. This is in contrast to most studies that have found serotypes 1/2a and 1/2b were to be the most common serotypes in food (Aarnisalo et al., 2003), a finding also supported by Gilot et al., (1996) in a study of foods in Belgium. A study by Wallace et al., (2003) also found serovar 1/2a in 90% of all the Listeria monocytogenes isolates tested in food samples. The results obtained from the current study indicated a correlation between certain serotypes and specific food products, serogroups 1/2b and 3b was absent from Cabbage, a food type that had more L. monocytogenes isolates than other food type. This observation is in line with results obtained by Vitas and Garcia-Jalon (2004) in a study of fresh and processed foods in Navarra, Spain. In the present study serogroup 4b, 4d and 4e was present in 30% isolates retrieved from the food. This is significant because among the Listeria monocytogenes serotypes, serotype 4b has been the number one serotype associated with human listeriosis (Zhang et al., 2007).

One major finding in this study is that the majority of the isolates found in food samples belonged to lineage III (4a, 4c), which is contrary to results found in previous studies (Gray et al., 2004; Ward et al., 2004). Isolates in lineage III are known to be more prevalent in animals with clinical Listeriosis (Jeffers et al., 2001). What was also significant was that isolates belonging to division II were not detected; no isolates were positive for PCR using D2 specific primers. Division two has serovar 1/2a, a serotype common in food products. Furthermore, four samples belonged to division I, namely; S10, S31, S200 and S258 from salads, V208 and V225 from cabbage, but could not be characterized further into serogroups because they proved to be negative for PCR serogroup identification using GLT and ORF2110 specific primers. This proved to be one of the major limitation of serotype identification by PCR, in that some isolates could not be conclusively be allocated to serogroups. One other shortcoming of serotyping by PCR is that isolates could not definitively be allocated to specific serotypes but only indicated a number of possible serogroups or a division. However, these results are not surprising as the PCR assays used were not based on genes encoding serotype-specific antigens.

Using REP-PCR Typing a significant observation in the results was that most diverse isolates in this study were more common in cabbage and salads than the dairy products. This was to be expected as relatively few Listeria isolates were isolated from dairy products. Variations within the sizes of PCR generated fragments using REP-PCR was observed with this studies and two studies carried out by Wojciech et al., (2004). The amplicons obtained by Wojciech et al., (2004) were shorter with sizes ranging between 123 to 735 bp, in comparison to amplicons obtained by Jersek et al., (1998) with sizes ranging from 298 to 6100 bp. In this study amplicons obtained ranged from 200bp to 3000 bp. Though the same primers were used, the reason of such differences could be the variation in the DNA polymerase used, PCR machines and protocols used in this study.

In this study, REP-PCR was used as a tool to characterize L. monocytogenes strains isolated from food. This method showed great possibilities for the typing of L. monocytogenes, as isolated strains showed very similar REP-PCR fingerprints by visual comparison and cluster analysis but managed to show diversity between strains. These data supports previous studies that suggest that REP-PCR can be used as an alternative method for typing *L. monocytogenes* such as automated ribotyping which has proven to be a valuable epidemiological tool in investigations of many infectious diseases.

5.1 Conclusion

The findings clearly highlight the possible occurrence of *L. monocytogenes* 1/2b and 4b among foods served by retailers and street vendors in Gaborone. The presence of this human pathogen in ready-to-eat foods should be considered as having significant public health implications, particularly among the immuno-compromised and HIV/AIDS persons who are at greater risk. In order to solve the problems in this study, there is a need for close co-operation between the products suppliers, supermarket management, workers, especially cleaning agents, the staff of cleaning companies, and hygiene specialists from the food industry and health inspectors from the Ministry of health. Street vendors need to be educated about ill health imposed by consumption of food contaminated by *L. monocytogenes*, on the other hand retailers need to be trained and be vigilant of retail assistants on the importance of maintaining the cold chain in order to prevent food borne disease outbreaks. Considering the occurrence of power disruptions that occur now and then and the high ambient temperatures experienced, Supermarket managers should be encouraged to invest in stand-by generators to serve during periods when power cuts occur.

6. Acknowledgements

The authors wish to convey sincere gratitude and thanks to the University of Botswana, University of South Africa, Walter Sisulu University and National Research Foundation (NRF) South Africa, for providing financial assistance for this research work. We are indebted to Mr Daniel Loeto for the great technical assistance provided during this work. Special thanks go to Dr M. Ditlhogo, Ex-Head of Biological Sciences Department and all the microbiology staff for their unique roles. Finally, appreciation is extended to supermarket managers and street vendors for affording us the opportunity to collect the various samples.

7. References

Aguado, V., Vitas, A. I. and Garcia-Jalon, I. 2004. Characterization of *Listeria monocytogenes* and *Listeria innocua* from vegetable processing plant by RAPD and ERA. *Internationasl Journal of Food Micrbiology* 90: 341-347.

Bille, J., Catinel, B., Bannerman, E., Jacquet, C., Yersin, M.N., Caniaux, I., Monget, D. and Rocourt, J. 1992. API *Listeria*, a new and promising one-day system to identify *Listeria* isolates. *Journal of Applied and Environmental Microbiology* 58: 1857-1860.

Baloga, C., Bernande, C., Passone, N., Minssart, P., Kamalo, C., Mbolidi, D. and Germani, Y. 2006. Primary and opportunistic pathogens associated with meningitis in adults in Benguin, Central African Republic. In relation to human immunodeficiency virus serostatus. *International Journal ofInfectious Diseases* 10: 387-395.

Boruki, M. K. and Call, D. R., 2003. *Listeria monocytogenes* Serotype Identification by PCR. *Journal of Clinical Microbiology*, Vol. 41, No. 12 (December 2003), pp. (5537-5540), ISBN: 0095-1137.

Buchanan, R.L, Smith, J.L, and Long W. 2000. Microbial risk assessment: dose-response relations and risk characterization. International *Journal of Food Microbiology*, Vol. 58, No. 2000 (November 2000), pp. (159-172), ISBN: PIIS0168-1605(00)00270-1.

Charpentier, E., and Courvalin, P., 1999. Antibiotic resistance in *Listeria* species. *American Society for Microbiology*, Vol. 43, No. 1 (July 1995), pp. 2103-2108, ISBN: 0022-1800.

Charpentier, E., Gerbaud, G., Jacquet, C., Rocourt, J. and Courvalin, P. 1995. Incidence of antibiotic resistance in *Listeria* species. *Journal of Infectious Diseases* 172: 277-281.

Chen, Y. and Knabel, S. J., 2007. Multiplex PCR for simultaneous detection of bacteria of the genus Listeria, *Listeria monocytogenes* and Major serotypes and epidemic clones of *L. monocytogenes*. *Applied and Environmental Microbiology*, Vol. 73, No. 19 (August 2007), pp. 6299-6304, ISBN: 0099-2240.

Chintu C., Bathirunathan N., 1975. Bacterial meningitis in infancy and childhood in Lusaka (One year prospective study). *Medical Journal of Zambia*.9:150-7

Chopra I.,and Roberts M. 2001. Tetracycline Antibiotics: Mode of action, Application, Molecular Biology, and Epidemiology of Bacterial Resistance. MMBR 65: 232-260.

Doumith, M., Buchrieser, C., Glaser, P., Jacquet, C., and Martin, P., 2004. Differentiation of major *Listeria monocytogenes serovars by multiplex PCR. Journal of Clinical Microbiology, Vol.42 No.8 (August 2004), pp. (3819-38220), SBN 0095-1137.*

Duffy, L.L., Vanderlinde, P.B., and Grau, F.H. 1994. Growth of *Listeria monocytogenes* on vacuum-packed cooked meats; effects of low pH, aw, nitrite and ascorbate. *International Journal of Food Microbiology*, Vol. 23, issues 3-4 (November 1994), pp. (377-390), ISSN:1517-8382.

Farber, J. M, and Peterkin, P.I., 1991. *Listeria monocytogenes*: a food-borne pathogen. *Microbiology Review*, Vol. 55, No. 4 (September 1991), pp. (476-511), ISBN: 04448-814981.

Gilot, P., Genicot, A., and Andre, P. 1996. Serotyping and esterase typing for analysis of *Listeria monocytogenes* populations recovered from foodstuffs and from human patients with listeriosis in Belgium. *Journal of Clinical Microbiology* 34: 1007-1010.

Gray, M. J., Zadoks, R. N., Fortes, E. D. and 7 other authors (2004). *Listeria monocytogenes isolates from foods and humans form distinct but overlapping populations. Applied and Environmental Microbiology* 70, 5833–5841.

Hohne K., LooseB., Seeliger H.P.,1975. Isolation of *Listeria monocytogenes* in slaughter animals and bats of Togo (West Africa).*Annual Microbiology (Paris)*.126A:501-7.

Leimeister-Wachter M. , and Chakraborty T . 1989. Detection of Listeriolysin , the Thiol – Dependent Hemolysin in *Listeria monocytogenes* , *Listeria ivanovii* , and *Listeria seeligeri* . *America Society for Microbiology* 57: 2350-2357.

Huss, H.H., Jorgenssen, L.V., and Vogel, B.F. 2000. Control options for *Listeria monocytogenes* in seafood. *International Journal of Food Microbiology*, Vol. 62, No. 1 (November 2000), pp. (267-274), ISBN:0021-9193.

Jacquet, C., Gouin, E., Jeannel, D., Cossart, P. and Rocourt. J., 2002. Expression of ActA, Ami, InlB, and listeriolysin O in *Listeria monocytogenes* of human and food origin. *Applied Environmental Microbiology*, Vol. 68, No. 1 (November 2002), pp. (616-622).

Jeffers, G. T., Bruce, J. L., McDonough, P. L., Scarlett, J., Boor, K. J. and Wiedmann, M. (2001). Comparative genetic characterization of *Listeria monocytogenes* isolates from human and animal listeriosis cases. *Microbiology* 147, 1095–1104.

Jersek B. P., Gilot M., Klun N. and Mehla J., 1999. Typing of Listeria monocytogenes strains based on repetitive element sequences-based PCR. *Journal of Clinical Microbiology.* 37:103-109.

Kathariou, S. 2000. Pathogenesis determinants of *Listeria monocytogenes,* p. 295-314. In J.W. Cary, J.E. Linz and D. Bhatnagar (ed.), Microbial foodborne diseases. Technomics Publishing Co., Inc., Lactaster, PA.

Kunene, N.F., J.W. Hastings and A. Von Holy, 1999. Bacterial populations associated with a sorghum based fermented weaning cereal. *International Journal of Food Microbiology* 49(1-2): 75-83.

Kurunasagar, I., Krohne, G. and Goebel W., 1993. *Listeria ivanovii* is capable of cell-to-cell spread involving Actin Polymerization. *Journal of Infection and Immunity,* Vol. 61, No.1 (October 1993), pp. (162-169), ISBN: 0019-9567.

Manani, T. A., Collison, E. K., Mpuchane, S. 2006. Microflora of minimally processed frozen vegetables sold in Botswana. *Journal of Food Protection,* Vol. 69, No. 11 (November 2006), pp. (2581 – 2586), ISBN: 17133799 Pubmed.

McLauchlin, J., Mitchel, R.T., Smerdon, W.J., and Jewell, K., 2004. *Listeria monocytogenes* and listeriosis. A review of hazard characterization for use in microbial risk assessment of foods. *International Journal of Food Microbiology,* Vol. 92, No. 3 (October 2004), pp. (15-33), ISBN: 0387890254.

Mead, P.S., Slutsker, L., Dietz, V., McCraig, L.F., Breese, J.S., Shapiro, C., Griffin, P.M., and Tauxe, R.V., 1999. Food-related illness and death in the United States. *Emerging Infectious Diseases,* Vol. 5, No.5 (October 1999), pp. (607-62), ISBN: 10511517 Pubmed.

Mosupye F. M. and Von Holy A. 2000. Microbiological hazard identification and exposure assessment of street food vending in Johannesburg, South Africa. *International Journal of Food Microbiology* 61:137-145.

Njagi L.W., Mbathia P.G., Bebora L.C., Nyaga P.N., Minga U., Olsen J. E., 2004. Carrier status for *Listeria monocytogenes* and other *Listeria* species in free range farm and market healthy indigenous chicken and ducks.*East African Medical Journal.*81:529-33

Pitcher, D.G., Saunders, N.A. and Owen R.J., 1989. Rapid extraction of bacterial genomic DNA with guanidium thiocyanate. *Letters in Applied Microbiology,* Vol. 8, No. 1 (January 1989), pp. (151-156), ISBN: 1472765X.

Poyart-Salmeron, C., Trieu-Cuot, P., Carlier, C., MacGowan, A., McLauchlin J., and Courvalin, P. 1992. Genetic basis of tetracycline resistance in clinical isolates of *Listeria monocytogenes, Antimicrobial Agents and Chemotherapy* 36: 463–466.

Rodriguez-Lazaro, D., Hernandez, M., Scortti, M., Esteve, T., Vazquez-Boland, J.A. and Pla, M. 2004. Quantitative detection of *Listeria monocytogenes* and *Listeria innocua* by Real-Time PCR: Assessment of *hly, iap* and lin 02483 targets and Amplifluor technology. *Journal of Applied and Environmental Microbiology* 70: 1366-1377.

Rodriguez, L., and Cepeda Saez, A. 1994. Susceptibilities of *Listeria* species isolated from food to nine antimicrobial agents, *Antimicrobial Agents and Chemotherapy,* Vol. 38, No. 7 (July 1994), pp. (1655–1657), ISBN:0066-4804.

Seeliger, H.P. and Hohne, K. 1979. Serotyping of *Listeria monocytogenes* and related species. *Methods in Microbiology* 13: 31-49.

Wallace, F.M., Call, JE., Porto, A.C.S., Cocoma, G.J., ERRC Special Projects Team, and Luchansky., J.B. 2003. Recovery rate of *Listeria monocytogenes* from commercially prepared frankfurters during extended refrigerated storage. *Journal of Food Protection*, Vol. 66, No.1 (December 2003), pp. (584-591), ISBN:0781793947.

Ward, T. J., Gorski, L., Borucki, M. K., Mandrell, R. E., Hutchins, J. and Pupedis, K. (2004). Intraspecific phylogeny and lineage group identification based on the *prfA* virulence gene cluster of *Listeria monocytogenes*. *Journal of Bacteriology* 186, 4994–5002.

Wiedmann, M., Barany, F., and Batt, C.A., 1993. Detection of *Listeria monocytogenes* with a Nonisotopic Polymerase Chain Reaction-Coupled Ligase Chain Reaction Assay. *Applied Enviromental Microbiology*, Vol. 59, No.8 (August 2003), pp. (2743-2745), ISBN:0099-2240.

Wojciech, L., Kowalczyk, K., Staroniewicz, Z., Kosek, K., Molenda, J., and Ugorski, M. 2004.Genotypic characterization of *Listeria monocytogenes* from foodstuffs and farm animals in Poland. *Journal of Bacteriology*, Vol. 186, No.14 (July 2004), pp. (427-435), ISBN:1740-8261.

Wong, A.C. 1998. Biofilms in food processing environment. *Journal of Dairy Science* 81: 2765-2770.

Wouters, P. C., Dutreux, N., Smelt, J. P. P., and Lelieveld, H. L. M., 1999. Effects of Pulse Electric Fields on Inactivation Kinetics of *Listeria innocua*. *Applied and Enviromental Microbiology*, VOL. 65, No. 12 (September 1999), pp. (5364-5371), ISNB:0099-2240.

Yöcel, N., Citak, S. and Önder, M., 2005. Prevalence and resistance of Listeria species in meat products in Ankara, Turkey. *Food Microbiology*, Vol. 22, No. 2005 (March 2005), pp. (241-245), ISBN:0740-0020.

Zang, W. and Knabel, S.J. 2005. Multiplex PCR assay simplifies serotyping and sequence typing of *Listeria monocytogenes* associated with human outbreaks. *Journal of Food Protection* 68: 1907-1910.

Zhang, Y., Yeh, E., Hall, G., Cripe, J., Bhagwat, A.A. 2007. Characterization of *Listeria monocytogenes* isolated from retail foods. *International Journal of Food Microbiology*, Vol. 113, No. 2007 (July 2006), pp. (47-53), ISBN:1845938801.

8

Extending the Shelf Life of Fresh Marula (*Sclerocarya birrea*) Juice by Altering Its Physico-Chemical Parameters

S. Dube[1], N.R. Dlamini[2], I. Shereni[1] and T. Sibanda[1]

[1]*Department of Applied Biology and Biochemistry,*
National University of Science and Technology, Ascot, Bulawayo,
[2]*CSIR Biosciences, Pretoria*
[1]*Zimbabwe*
[2]*South Africa*

1. Introduction

1.1 Taxonomy and distribution

Sclerocarya birrea (A. Rich.) Hochst. subsp. *caffra* (Sond.) (marula) is a very common and widespread throughout much of sub-Saharan Africa and is a member of the Anacardiaceae family, along with 650 species and 70 genera of mainly tropical or subtropical evergreen or deciduous trees, shrubs and woody vines along side with the mango (*Mangifera indica* L.). *S. birrea* has three recognised subspecies; fruit-bearing species of which *Sclerocarya birrea* subsp. *caffra* is the most ubiquitous and occurs in east tropical Africa (Kenya, Tanzania), south tropical Africa (Angola, Malawi, Mozambique, Zambia and Zimbabwe) and southern Africa (Botswana, Namibia, South Africa and Swaziland) as well as Madagascar. *Sclerocarya birrea* subsp. *multifoliolata* (Engl.) Kokwaro subspecies occurs in mixed deciduous woodland and wooded grassland in Tanzania. The tree grows in a wide variety of soils but prefers well-drained soil. It exists at altitudes varying from sea level to 1800 m and an annual rainfall range of 200 – 1500 mm. The key factor limiting its distribution is its sensitivity to frost (Von, 1983; Palgrave, 1990; Mizrahi and Nerd, 1996; Aganga and Mosase, 2001; Emanuel, *et al.*, 2005).

1.2 Flowering and fruiting

S. birrea is a dioecious species. Male and female flowers occur separately, but not always, on separate trees. The flowers are small, with red sepals and yellow petals, and are borne in 50 – 80 mm-long sprays of small oblong clusters (Venter and Venter, 1996; van Wyk and van Wyk, 1997; van Wyk *et al.*, 2002).

1.3 Traditional uses of marula fruit

It has many uses, including the fruits that are eaten fresh or after fermentation to make a beer, the kernels are eaten or the oil extracted, the leaves are browsed by animals and have medicinal properties, as does the bark. The wood is carved into traditional utensils such as spoons and plates as well as decorative curios. Because of the widespread occurrence and use of *S. birrea* it has frequently been identified as a key species in the development of rural businesses using the fruit, beer, nuts or oil and therefore a good candidate for domestication (Fox and Norwood, 1982; Nerd and Mizrahi, 1993; Mizrahi and Nerd, 1996; Shackleton *et al.*, 2002). The fruit and nuts have extensive commercial value. The fruit is described as having an exotic flavor and high nutritive value (for example, vitamin C content is up to 3 times higher in 'marula' fruit than equivalent weight of oranges. The fruit is often fermented to give an alcoholic beverage. Some tribes, such as the Pedi, prepare a relish from the leaves of *S. birrea*. The Zulu people of South Africa regard *S. birrea* fruit as a potent insecticide (van Wyk, *et al.*, 2002; Mdluli, 2005). The whole fruit is used in many parts of southern Africa for brewing beer and distilling spirits. In Mozambique, South Africa and Zambia, the fruit of *S. birrea* is used to flavor liqueur. The gum obtained from the tree is rich in tannins, and hence, it is employed in the production of an ink substitute. Zimbabwean and South African villagers benefit from the customary usage of *S. birrea* wood in the manufacture of dishes, maize stamping mortars, drums, toys, curios, divining bowls and carvings. The peel from *S. birrea* fruit is of paramount importance in the production of oil for cosmetic purposes (Ojewole, *et al.*, 2010).

1.4 The marula fruit

Female trees bear pale-yellow, round fruit with thick leather-like rind, and fibrous, juicy, mucilaginous flesh. The fruit, which is about the size of a golf ball, is pale-yellow when ripe, with a diameter of 30 – 40 mm (Nerd and Mizrahi, 1993; Ojewole *et al.*, 2010). The fruit has sour taste, and is much sought-after by birds, game mammals and humans, because of its delicious pulp, and edible, tasty nuts (van Wyk, *et al.*, 2002). The fruit pulp contains citric acid, vitamin C and sugar, while the nut is rich in non-drying oil and protein (van Wyk, *et al.*, 2002; Hillman, *et al.*, 2008; Ojewole, *et al.*, 2010). The only way to make this fruit juice available to children before fermentation and is by finding ways to extend its shelf life. The average *Sclerocarya birrea* (marula) tree under the climatic conditions of Matebeleland South produces 20000±5000 fruits per season. The fruits ripen between February and April. The time from ripening onset to rotting of the fruit was 16±4 days. Skin colour starts to change on day 3 after abscission, and a completely yellow colour is obtained by day 12.

1.5 Fruit ripening

Marula fruits abscise before ripening; at this stage the skin color is green and the fruit is firm. The time for fruit abscission varies among trees. Of the studied trees, fruits abscised mainly in March and in late April. This can be attributed to genetic variation, which can be exploited for expanding the harvest period by planting clones that ripen at different times. (Palgrave, 1990; Mdluli, 2005; Dlamini and Dube, 2008).

2. Marula pulp characteristics

Of the wide range of nutrients in the Marula pulp the Vitamin C content has attracted the most attention. Indeed the vitamin C content is important to local communities who know well that it prevents scurvy. The fruit has small amounts of other vitamins such as thiamine, riboflavin and nicotinic acid. It is 85% moisture and 14% carbohydrate, mostly sucrose.

Citric acid is the most abundant acid excluding ascorbic acid, but malic and tartaric acid have also been noted. The mineral composition of the fruit shows high concentrations of Potassium, Calcium and Magnesium (Shackleton, et al., 2010). The aroma of Marula is perceived by other people to be like that of the grapefruit and there are in fact compounds that the two fruits have in common. However, the similarity in taste between grapefruit and marula is probably because of a dominant bitter taste caused by non volatiles. Others have likened the smell of marula juice to pineapples but this is probably also only in part due to complimentary volatile components such as ethyl acetate, benzaldehyde, linalool. With a unique and pleasurable taste, an attractive colour and odour and health properties such as high vitamin C and potassium, this fruit has all the physical requirements of the growing it for industrial purposes. Marula, *Scelerocarya birrea*, subspecies *caffera*, is one of Africa's botanical treasures. The flesh of the fruit clings onto to its brown stone. The flesh is very fibrous and juicy. Inside the woody stone are two to three seeds, which are very rich in oil and protein. (Shackleton, *et al.*, 2010).

Generally edible fruits constitute a commercially important and nutritionally indispensable food commodity. Being a part of a balanced diet, fruits play a vital role in human nutrition by supplying the necessary growth regulating factors essential for maintaining normal health. Fruits are widely distributed in nature. One of the limiting factors that influence their economic value is the relatively short ripening period and reduced post-harvest life. Fruit ripening is a highly coordinated, genetically programmed, and an irreversible phenomenon involving a series of physiological, biochemical, and organoleptic changes, that finally leads to the development of a soft edible ripe fruit with desirable quality attributes. Excessive textural softening during ripening leads to adverse effects/spoilage upon storage. Carbohydrates play a major role in the ripening process, by way of depolymerization leading to decreased molecular size with concomitant increase in the levels of ripening inducing specific enzymes, whose target differ from fruit to fruit. (Hillman, *et al* 2008; Brumell and Harpster, 2001; Prasanna, *et al.*, 2007; Prescott, 2011) The major classes of cell wall polysaccharides that undergo modifications during ripening are starch, pectins, cellulose, and hemicelluloses. Pectins are the common and major components of primary cell wall and middle lamella, contributing to the texture and quality of fruits. Their degradation during ripening seems to be responsible for tissue softening of a number of fruits. Structurally pectins are a diverse group of heteropolysaccharides containing partially methylated D-galacturonic acid residues with side chain appendages of several neutral polysaccharides. The degree of polymerization/esterification and the proportion of neutral sugar residues/side chains are the principal factors contributing to their (micro-) heterogeneity. Pectin degrading enzymes such as polygalacturonase, pectin methyl esterase, lyase, and rhamnogalacturonase are the most implicated in fruit-tissue softening. Recent advances in molecular biology have provided a better understanding of the biochemistry of fruit ripening as well as providing a hand for genetic manipulation of the entire ripening process. It is desirable that significant breakthroughs in such related areas will come forth in the near future, leading to considerable societal benefits(Brumell and Harpster, 2001; Prasanna, *et al.*, 2007; Prescott, 2011).

Excessive softening is the main factor limiting fruit shelf life and storage. Transgenic plants modified in the expression of cell wall modifying proteins have been used to investigate the role of particular activities in fruit softening during ripening, and in the manufacture of processed fruit products. Transgenic experiments show that polygalacturonase (PG) activity is largely responsible for pectin depolymerization and solubilization, but that PG-mediated pectin depolymerization requires pectin to be de-methyl-esterified by pectin methylesterase (PME), and that the PG beta-subunit protein plays a role in limiting pectin solubilization. Suppression of PG activity only slightly reduces fruit softening (but extends fruit shelf life), suppression of PME activity does not affect firmness during normal ripening, and suppression of beta-subunit protein accumulation increases softening. All these pectin-modifying proteins affect the integrity of the middle lamella, which controls cell-to-cell adhesion and thus influences fruit texture (Arthachinta, 2000; Brumell and Harpster, 2001; Prasanna, *et al.*, 2007; Prescott, 2011). Diminished accumulation of either PG or PME activity considerably increases the viscosity of tomato juice or paste, which is correlated with reduced polyuronide depolymerization during processing. In contrast, suppression of beta-galactosidase activity early in ripening significantly reduces fruit softening, suggesting that the removal of pectic galactan side-chains is an important factor in the cell wall changes leading to ripening-related firmness loss. (Brumell and Harpster, 2001; Prasanna, *et al.*, 2007; Prescott, 2011). Suppression or over expression of endo-(1--4) beta-D-glucanase activity has no detectable effect on fruit softening or the depolymerization of matrix glycans, and neither the substrate nor the function for this enzyme has been determined. The role of xyloglucan endotransglycosylase activity in softening is also obscure, and the activity responsible for xyloglucan depolymerization during ripening, a major contributor to softening, has not yet been identified. However, ripening-related expansion protein abundance is directly correlated with fruit softening and has additional indirect effects on pectin depolymerization, showing that this protein is intimately involved in the softening process. (Brumell and Harpster, 2001; Prasanna, *et al.*, 2007; Prescott, 2011). Transgenic work has shown that the cell wall changes leading to fruit softening and textural changes are complex, and involve the coordinated and interdependent activities of a range of cell wall-modifying proteins. It is suggested that the cell wall changes caused early in ripening by the activities of some enzymes, notably beta-galactosidase and ripening-related expansin, may restrict or control the activities of other ripening-related enzymes necessary for the fruit softening process (Brumell and Harpster, 2001; Prasanna *et al.*, 2007; Prescott, 2011). Whether the sugar of fruits is formed within them, or introduced through the stem, and, if formed in the fruits, from what substance formed, are questions which have been investigated, but not wholly settled. It has been pretty generally held that starch in the unripe fruits is converted into sugar in the ripe fruits; the fruit acids inducing the change, as we know they have power to do. But starch is not found in the unripe stage of all fruits, and, in the cases where found, its quantity is sometimes too small to serve as the source of all the sugar of the ripened fruit The maturity of fruit is the period of its maximum quantity of sugar. Sooner or later, the quantity of sugar begins to diminish, and then the fruit is overripe. It is safe to say that the sugar often begins to decompose during the life of the fruit; that is to say, fruit becomes overripe during its life. It would be difficult, however, to fix on the termination of the life of fruit. We certainly cannot say that life ceases when the circulation with the plant is cut off; and we cannot say that life continues in the sarcocarp until it is wholly disintegrated. (Brumell and Harpster, 2001; Prasanna, *et al.*, 2007; Prescott, 2011). Now it is within the limits of our subject to inquire by what changes the sugar begins to disappear. In general terms, sugar suffers oxidation in ripe fruits, small portions being oxidized away even during the production of larger portions, and before perfect maturity. We do not know what fruit constituents, if any, result in this oxidation. The final products of oxidation, carbonic acid and water, are exhaled during ripening, and with greater rapidity after maturity has been passed. The quantity of acids in fruits usually diminishes during ripening. The diminution is not, however, nearly so great as it appears to the taste, because the acid of ripe fruits is masked to the taste by the larger proportions of sugar and the pectous substances then present. The removal of acids is chiefly due to oxidation. It is not found that acids are neutralized, to any considerable extent, during ripening, by alkalies conveyed through the stem (Brumell and Harpster, 2001; Prasanna *et al.*, 2007; Prescott, 2011). It is stated that the acids continue to oxidize away, after the sugar has reached its maximum and before it begins to diminish. Hence, perfect ripeness in fruit has been defined as that period during the maximum quantity of sugar when the quantity of acid is least. This will be, of course, just before the sugar begins to diminish. It has been stated that both citric and malic acids are often found in unripe grapes, and are substituted by tartaric acid during the ripening. Oxalic acid is more often found in unripe than in ripe fruits. It is to be desired that closer determinations should be made as to the presence and proportion of oxalic acid in tomatoes and some other fruits (Brumell and Harpster, 2001; Prasanna *et al.*, 2007; Prescott, 2011).

3. Fresh marula fruit juice shelf life extension by high-pressure processing – pascalization and aseptic methods

The application of high hydrostatic pressure in processing of food is of great interest because of its ability to inactivate food related microorganisms and enzymes, at low temperature, without the need for chemical preservatives. High hydrostatic pressures, around 650 MPa reduces the microbial load in foods such as fruits. Pressure-treated foods have organoleptic properties similar to fresh products, which is a major advantage in juice processing as it matches consumer demand for healthy, nutritious and "natural" products. However, an important issue arises when we consider the acceptance of such products by the consumer (Jay, 1986; Adams and Moss, 1995; Deliza *et al.*, 2004). It is also noted that the microbicidal activity of high pressure is enhanced by low pH or temperatures above and below ambient. High hydrostatic pressure acts primarily on non-covalent linkages, such as ionic bonds, hydrogen bonds and hydrophobic interactions, and it promotes reactions in which there is an overall decrease in volume (Adams and Moss, 1995; Indrawati *et al.*, 2008; Sampedro *et al.*, 2008; Chao, *et al* 2011). It can have profound effects on proteins, where such interactions are critical to structure and function, although the effect is variable and depends on individual protein structure. Other proteins are relatively unaffected and this can cause problems when they have enzymic activity which limits product shelf-life. Pectin esterase in orange juice, for instance, must be inactivated to stabilize the desired product cloudiness, but is very stable to pressures up to 1000 MPa. Non-protein macromolecules can also be affected by high pressures so that pascalized starch products often taste sweeter due to conformational changes in the starch which allow salivary amylase greater access. Adverse effects on protein structure and activity obviously contribute to the antimicrobial effect of high pressures, although the cell membrane also appears to be an important target. Membrane lipid bilayers have been shown to compress under pressure and this alters their permeability. As a general rule vegetative bacteria and fungi can be reduced by at least one log cycle by 400 MPa applied for 5 min. (Indrawati *et al.*, 2008; Sampedro *et al.*, 2008; Chao *et al.*, 2011)

Bacterial endospores are more resistant to hydrostatic pressure, tolerating pressures as high as 1200 MPa. Their susceptibility can be increased considerably by modest increases in temperature, when quite low pressures (100 MPa) can produce spore germination, a process in which the spores lose their resistance to heat and to elevated pressure. Hydrostatic processing has a number of appealing features for the food preparations such as Marula fruit juice. (Indrawati, *et al.*, 2008; Sampedro, *et al.*, 2008; Chao, *et al.*, 2011)

It acts instantly and uniformly throughout a substance so that the processing time is not related to container size and there is none of the penetration problems associated with heat processing. Nutritional quality, flavour, appearance and texture resemble the fresh material very closely. To the consumer where it has been used it is regarded as a 'natural' process with none of the negative associations of processes such as irradiation or chemical preservatives. ((Indrawati, *et al.*, 2008; Sampedro, *et al.*, 2008; Chao, *et al.*, 2011)

Appertized foods include those which are hermetically sealed into containers, usually cans, and then subjected to heat process in-pack. While this has been hugely successful as a long term method of food preservation, it does require extended heating periods in which a food's functional and chemical properties can be adversely affected. (Adams and Moss, 1995)

3.1 Fresh marula fruit juice shelf life extension by temperature control

Low-temperature storage - chilling and freezing the rates of most chemical reactions are temperature dependent; as the temperature is lowered so the rate decreases. Since food spoilage is usually a result of chemical reactions mediated by microbial and endogenous enzymes, the useful life of many foods can be increased by storage at low temperatures. Chilled foods are those foods stored at temperatures near, but above their freezing point, typically 0-5°C. Chill storage can change both the nature of spoilage and the rate at which it occurs. There may be qualitative changes in spoilage characteristics as low temperatures exert a selective effect preventing the growth of mesophiles and leading to a microflora dominated by psychrotrophs (Paine and Paine, 1992; Blakestone, 1999; Cheikh, *et al.*, 2009; Philip, 2010).

Blanching is achieved either by brief immersion of foods into hot water or the use of steam. Its primary functions are as follows: inactivation of enzymes that might cause undesirable changes during freezing storage, enhancement or fixing of the green color of certain vegetables, reduction in the numbers of microorganisms on the foods, facilitating the packing of leafy vegetables by inducing wilting, displacement of entrapped air in the plant tissues. When water is used, it is important that bacterial spores not be allowed to build up sufficiently to contaminate the juice. Reductions of initial microbial loads as high as 99% have been claimed upon blanching (Blakestone, 1999; Cheikh, *et al.*, 2009; Philip 2010).

According to Adams and Moss,1995 the term Pasteurization is given to heat processes typically in the range 60-80 °C and applied for up to a few minutes, is used for two purposes. First is the elimination of a specific pathogen or pathogens associated with a product. This type of pasteurization is often a legal requirement introduced as a public health measure when a product has been frequently implicated as a vehicle of illness. The second reason for pasteurizing a product is to eliminate a large proportion of potential spoilage organisms, thus extending its shelf-life. This is normally the objective when acidic products such as beers, fruit juices, pickles, and sauces are pasteurized. Where pasteurization is introduced to improve safety, its effect can be doubly beneficial (Adams and Moss, 1995; Blakestone, 1999; Cheikh, *et al.*, 2009; Philip, 2010). The process cannot discriminate between the target pathogen(s) and other organisms with similar heat sensitivity so a pasteurization which destroys say *Salmonella* will also improve shelf-life. The converse does not normally apply since products pasteurized to improve keeping quality are often intrinsically safe due to other factors such as low pH. On its own, the contribution of pasteurization to extension of shelf-life can be quite small; particularly if the pasteurized food lacks other contributing preservative factors such as low pH, where as appertization refers to processes where the only organisms that survive processing are non-pathogenic and incapable of developing within the product under normal conditions of storage. As a result, appertized products have a long shelf-life even when stored at ambient temperatures. An appertized or commercially sterile food is not necessarily sterile - completely free from viable organisms (Adams and Moss, 1995; Blakestone, 1999; Cheikh, *et al.*, 2009; Philip, 2010).

Sterilization: This is the process of destroying all forms of microbial life. A sterile object is free from living organism. killing bacteria is the irreversible loss of the bacteria's ability to reproduce. The cells are killed over a period of time at a constant exponential rate that is the inverse of exponential growth rate. Some portion of population dies during any given time. The graph of logarithm of number of survivor's v/s time in hours shows that the death rate is constant. Slope of this curve is a measure of death-rate (Jay, 1986; Adams and Moss, 1995). The probability of killing the organisms is also proportional to concentration of chemical agent or intensity of physical agent. It takes time to kill the population and if we have many cells, we must treat them for a longer time. Although imperfect, cooking and canning are the most common applications of heat sterilization. Boiling water kills the vegetative stage of all common microbes. Cooking food does not sterilize food but simply reduces the number of disease-causing micro-organisms to a level that is not dangerous for people with normal digestive and immune systems. Pressure cooking is analogous to autoclaving and when performed correctly renders food sterile. (Jay, 1986; Adams and Moss, 1995).

In UHT processing the food is heat processed before it is packed and then sealed into sterilized containers in a sterile environment. This approach allows more rapid heating of the product, the use of higher temperatures than those employed in canning, typically 130-140°C, and processing times of seconds rather than minutes (Jay, 1986; Adams and Moss, 1995; Michael and Hepell 2000).The advantage of using higher temperatures is that the z value for chemical reactions such as vitamin loss, browning reactions and enzyme inactivation is typically 25-40°C compared with 10°C for spore inactivation. This means that they are less temperature sensitive so that higher temperatures will increase the microbial death rate more than they increase the loss of food quality associated with thermal reactions. A common packing system used in conjunction with UHT processing is a form/ fill/seal operation in which the container is formed in the packaging machine from a reel of plastic or laminate material, although some systems use preformed containers. In order to obtain commercial sterility it is given a bactericidal treatment, usually with hydrogen peroxide, sometimes coupled with UV irradiation (Jay, 1986; Adams and Moss, 1995).

3.2 Fresh marula fruit juice shelf life extension by chemical methods

Chemicals that can possibly be used for sterilization marula juice include the gases ethylene oxide, Ozone, hydrogen peroxide which are examples of chemical sterilization techniques based on oxidative capabilities.

Ethylene oxide (ETO) is the most commonly used form of chemical sterilization. Due to its low boiling point of 10.4°C at atmospheric pressure, Ethylene oxide behaves as a gas at room temperature. Ethylene oxide chemically reacts with amino acids, proteins, and DNA to prevent microbial reproduction. The sterilization process is carried out in a specialized gas chamber. After sterilization, products are transferred to an aeration cell, where they remain until the gas disperses and the product is safe to handle. Ethylene oxide is used for cellulose and plastics irradiation, usually in hermetically sealed packages. Ethylene oxide can be used with a wide range of plastics and other materials without affecting their integrity. Ethylene oxide vapours are inflammable so a mixture of ethylene oxide 10 to 20 % with 80 to 90 % CO_2, or Feron is used, CO_2 or Feron serve as inert diluent which prevent inflammability. It is a unique and powerful sterilizing agent. Bacterial spores show little resistance to destruction by this agent. (Morga, *et al.*, 1979; Adams and Moss, 1995) It has got very good penetrating power. It passes through the sterilizes large packets of materials and even certain plastics. It should be used with caution. The concentration of ethylene oxide and temperature and humidity are critical factors which determine the time required for sterilization. The apparatus used for its application is an autoclave modified. It is effective at low temperature and does not damage the material exposed to it but it is slow in action. The mode of action is believed to be alkylation reactions with organic compounds such enzymes and other proteins. (Jay, 1986; Adams and Moss, 1995). Ozone sterilization has been recently approved for use in the U.S. It uses oxygen that is subjected to an intense electrical field that separates oxygen molecules into atomic oxygen, which then combines with other oxygen molecules to form ozone. Ozone is used as a disinfectant for water and food. It is used in both gas and liquid forms as an antimicrobial agent in the treatment, storage and processing of foods (Adams and Moss, 1995; Cullen *et al.*, 2010; Philip, 2010).

Low Temperature Gas Plasma (LTGP) is used as an alternative to ethylene oxide. It uses a small amount of liquid hydrogen peroxide (H_2O_2), which is energized with radio frequency waves into gas plasma. This leads to the generation of free radicals and other chemical species, which destroy organisms (Jay, 1986; Adams and Moss, 1995).

Hypochlorites: Calcium hypochlorite $Ca(OCl)_2$ and sodium hypochlorite $NaOCl$ are widely used. They are in powder or liquid forms and in various concentrations from 5-70%. $CaCl_2$ is used to sanitize equipment and 1% $NaOCl$ for personal hygiene and household disinfection. (Adams and Moss, 1995; Cullen *et al.*, 2010; Philip, 2010). This could find application in the extension of marula juice shelf life extension.

3.3 Preservatives

Although some would regard all chemical additions to food as synonymous with adulteration, many are recognized as useful and are allowed. Additives may be used to aid processing, to modify a food's texture, flavour, nutritional quality or colour but, here, we are concerned with those which primarily effect keeping quality: preservatives(Jay, 1986). Preservatives are defined as 'substances capable of inhibiting, retarding or arresting the growth of micro-organisms or of any deterioration resulting from their presence or of masking the evidence of any such deterioration'. They do not therefore include substances which act by inhibiting a chemical reaction which can limit shelf-life, such as the control of rancidity or oxidative discoloration by antioxidants. Neither does it include a number of food additives which are used primarily for other purposes. Preservatives may be microbicidal and kill the target organisms or they may be microbistatic in which case they simply prevent them growing. This is very often a dose-dependent feature; higher levels of antimicrobial proving lethal while the lower concentrations that are generally permitted in foods tend to be microbistatic (Jay, 1986). For this reason chemical preservatives are useful only in controlling low levels of contamination and are not a substitute for good hygiene practices, effect on flavour and on product pH, thus potentiating their own action by increasing the proportion of undissociated acid present.

Benzoic acid occurs naturally in cherry bark, cranberries, greengage plums, tea and anise but is prepared synthetically for food use. Its antimicrobial activity is principally in the undissociated form and since it is a relatively strong acid (pH4.19) it is effective only in acid foods. As a consequence, its practical use is to inhibit the growth of spoilage yeasts and moulds (Jay, 1986).

3.4 Fresh marula fruit juice shelf life extension by radiation methods

Gamma rays and X-rays which have ionizing energy enough to pull electrons away from molecules and ionize them. When such radiation passes through cells it creates free hydrogen and hydroxyl radical and some peroxides which in turn can cause different kinds of intracellular damage. U.V. light does not ionize; it is absorbed quite specifically by different chemical species that can engage in a variety of chemical reactions not possible for unexcited molecules. Organisms may be subjected to acoustic radiation (sound waves). Ionizing radiation is also used to sterilize biological materials. This method is called Cold Sterilization because ionizing radiations produced relatively little heat in the material being irradiated. Thus it is possible to sterilize heat-sensitive substances by radiations and such techniques are being developed in the food and pharmaceutical industries (Jay, 1986; Adams and Moss, 1995; Chao *et al.*, 2011).

Ultraviolet light has the bactericidal activity. Although the radiant energy of sunlight, is partly composed of UV light, most of the shorter wavelength of this are filtered by the earth's atmosphere is restricted. Many lamps are available which emit a high concentration of UV light in the most effective region. UV light has very little ability to penetrate matter. UV light is absorbed by many cellular materials but most significantly by the nucleic acids where it does the greatest damage. The absorption and subsequent reactions are predominantly in the pyrimidines of the nucleotide bases which result in killing of cells. Death of a population of UV-irradiated cells demonstrates log-linear kinetics (Jay, 1986; Adams and Moss, 1995; Chao *et al.*, 2011).

Similar to thermal death and, in an analogous way, D values can be determined. These give the dose required to produce a tenfold reduction in surviving numbers where the dose, expressed in ergs, is the product of the intensity of the radiation and the time for which it is applied.

Determination of UV D values is not usually a straightforward affair since the incident radiation can be absorbed by other medium components and has very low penetration. Passage through 5 cm of clear water will reduce the intensity of UV radiation by two-thirds. This effect increases with the concentration of solutes and suspended material so that in milk 90% of the incident energy will be absorbed by a layer only 0.1 mm thick. This low penetrability limits application of UV radiation in the food industry to disinfection of air and surfaces (Jay, 1986; Adams and Moss, 1995; Chao *et al.*, 2011).

Low-pressure mercury vapor discharge lamps are used: 80% of their UV emission is at a wavelength of 254 ohm which has 85% of the biological activity of 260 om. Wavelengths below 200nm are screened out by surrounding the lamp with an absorbent glass since these wavelengths are absorbed by oxygen in the air producing ozone which is harmful. The output of these lamps falls off over time and they need to be monitored regularly. Process water can be disinfected by UV; this avoids the risk of tainting sometimes associated with chlorination, although the treated water will not have the residual antimicrobial properties of chlorinated water. Process workers must also be protected from UV since the wavelengths used can cause burning of the skin and eye disorders. (Jay, 1986; Adams and Moss, 1995; Chao *et al.*, 2011).

3.5 Gamma-rays

These are high energy radiations emitted from radioactive isotopes such as 60 C. they are similar to X-rays but of shorter wavelength. They have great penetration power and are lethal to all life including microbes. They are used for sterilization of materials of considerable thickness or volume. (Jay, 1986; Adams and Moss, 1995).

Microwave Radiation The microwave region of the e.m. spectrum occupies frequencies between 109 Hz up to 1012 Hz and so has relatively low quantum energy. Unlike the other forms of radiation, microwaves act indirectly on micro-organisms through the generation of heat. When a food containing water is placed in a microwave field, the dipolar water molecules align themselves with the field. As the field reverses its polarity 2 or 5 x 1 09 times each second, depending on the frequency used, the water molecules are continually oscillating. This kinetic energy is transmitted to neighbouring molecules leading to a rapid rise in temperature throughout the product. Microwaves are generated using a magnetron (Jay, 1986; Adams and Moss, 1995). The principal problem associated with the domestic use of microwaves is non-uniform heating of foods, due to the presence of cold spots in the oven, and the non-uniform dielectric properties of the food. These can lead to cold spots in some microwaved foods and concern over the risks associated with consumption of inadequately heated meals has led to more explicit instructions on microwaveable foods. These often specify a tempering period after heating to allow the temperature to equilibrate. (Jay, 1986; Adams and Moss, 1995; Prescott, 2011).

3.6 Fresh marula fruit juice shelf life extension by hormonal application

Most fruit first become ripened because of the release of a hormone within the fruit and the plant called ethylene ($H_2C=CH_2$). Ethylene causes the breakdown which in itself causes a production of enzymes that break down the structures of the fruit (e.g. amylase, pictenase).The ethylene gas also destroys the green pigment of the fruit chlorophyll. But for any fruit to become ripened there needs to be a high enough concentration of ethylene around the fruit to even begin to ripen. The process of the fruit spoiling is just the decomposition of the fruit itself. That is mostly caused by the two enzymes previously mentioned -amylase and pectinase. The role that these enzymes play are like biological scissors. The structure of plants are made out of carbohydrates - mostly starches (Jay, 1986;

Adams and Moss, 1995; Allong *et al.*, 2001; Prescott, 2011). The amylase acts on the starches, which are made of amylose, which is just a long chain of glucose molecule. And what the amylase does is it cuts the bonds of the starch chain, so then the glucose molecules are free; and now the fruit tastes sweet. The other enzyme pectinase does almost the exact same thing as amylase, it cuts away the bonds of the pectin molecule so now we have a bunch of pectinic acid, and now the fruit is soft enough to eat. Pectin is extremely important for fruits/plants because it hold the cells together. Fruits become discolored due to chemical reactions as well. Like previously stated before, ethylene destroys chlorophyll. When most fruits are produced, they all initially contain chlorophyll. Fruits then become ripened due to other chemicals within the fruit and as well as chemicals outside of the fruit (Jay, 1986; Adams and Moss, 1995; Prescott, 2011).

3.7 Fresh marula fruit juice shelf life extension by appropriate hybrids or genetic manipulation

Through selection hybrid trees that are short and easy to hand or machine pick can be produced. This would minimise contamination from the soil and improve the shelf life of the juice. This has been achieved in other plants where trees were bred to achieve heights where hand or machine picking was practiced with ease (Bergh and Whitsell, 1962).

4. Laboratory work on some aspects of marula juice shelf life extension

The aim of this study was to determine some aspects of physico-chemical conditions that could contribute to the shelf life extension of fresh marula juice.

Fruit juice is defined as the fermentable but unfermented juice pressed or squeezed from the fruit excluding the peel. Fresh juice has a very short life after extraction from the whole fruit, due to enzyme or microbial actions unless it is rapidly processed and/or preserved. Steps to achieve aseptic processing involve minimizing the presence of microorganism in the environment and on the product without compromising the organoleptic quality of the product. Since all fruit juices provide an excellent medium for microbiological activity it becomes imperative that high standards of hygiene are observed from the on set. To extract the juice the fruits were washed with chlorinated water and rolled on a clean hard surface and then pieced with a toothpick and the juice then squeezed out into clean containers. The single strength juice from different fruits was, pooled, filtered and subjected to the following treatments in 100ml containers: Blanching; Pasteurization; Sterilization; and Blanching /Pasteurization and then preserved with 0.1% Sodium Benzoate, one portion stored at 4°C and the other portion at 22°C for 8 weeks as a way of extending the shelf life. The following parameters were assessed, sugar content of the berry, acidity, and the pH at the green stage of the fruit to fully ripe yellow fruits to map out strategies for shelf life extension. There amount of Vitamin C (Ascorbic acid) was assessed for the following treatments Blanching; Pasteurization; Sterilization; and Blanching /Pasteurization as an indicator of stability. The °Brix value and Browning index were determined. The sensory scores were determined to ensure that the treatments done do not result in an unpalatable product. Measurements of headspace were determined to evaluate its effect on the shelf life. Extending the life of fresh marula juice by altering its physico chemical properties were carried out in the laboratory as follows.

1. Blanching in which the fruit juice was subjected to a temperature of 100°C for 3 minutes;
2. Pasteurization in which the fruit juice was subjected to a temperature of 82°C for 15 seconds;
3. Sterilization in which the fruit juice was subjected to a temperature of 100°C for 5 minutes;
4. Blanching and Pasteurization in which the fruit juice was subjected to both treatments as stated above. Each experiment had two replications to determine statistically significant differences between samples.
5. The juice from each treatment was preserved with 0.1% Sodium Benzoate, one portion stored at 4°C and the other portion at 22°C for 8 weeks.

Other Quality parameters were also assessed

6. Brix value; The oBrix was determined on a Riechart Abbe Refractometer. Distilled water was placed on the platform and the readings adjusted to zero then a drop of sample was placed on the platform and the total soluble solids (mostly sugars) value or °Brix value were read and recorded.
7. Browning index; The browning index was determined. The brown pigments were extracted by diluting the juice 1:1 with 95% ethanol, filtering and measuring the absorbance at 420nm on a Milton Roy spectrophotometer.
8. Total titratable acidity (TTA %) For total titratable acidity 0.1NaOH was titrated against 1ml sample single strength juice containing a drop of 1% phenolphthalein indicator in 95% ethanol to a permanent end point pink colour at pH 8.1.
9. Vitamin C content in mg/100g and vitamin C determination was done by taking 1ml of sample and titrating it with redox dye 2, 6-dichlorophenolindophenol (DCPIP) to a permanent pink colour and then concentration calculations were done. The juice of marula fruit is known to be rich in vitamin C (Hillman et al., 2008; Dlamini and Dube, 2008). Hillman et al., (2008), reported vitamin C contents of marula clones ranging from 7 to 21 mg/g dry weight, amounts that were approximately 10 times higher than orange and pomegranate fruit juice. The high ascorbic acid content of marula makes it a cheap and accessible source of nutrients and antioxidants. The protection against disease provided by fruits and vegetables has been attributed to the various antioxidants contained in these foods. Because of its strong reducing properties, ascorbic acid acts a singlet oxygen quencher, thus reducing the damaging effect of free radicals which are implicated in the etiology of a number of diseases, including cancer and heart, vascular, and neurodegenerative diseases (Hillman et al., 2008).
10. Sensory score were monitored weekly for eight weeks Sensory score was done after previously documented methods. A trained 10-member panel holds the sample in the mouth, smell it and visualise it. The samples were coded with three-digit random numbers, placed in random order and served at 16°C in wine glasses. Unsalted crackers and water rinsing were utilized between each sample. Each panelist was asked to rate the samples for clarity, degree of browning, color, taste and flavor. A score sheet with vertical line scales for each rating was provided. Each scale was 10-cm long with anchor terms labeling each end. The top of the clarity scale was anchored with clear, and the bottom was anchored with extremely cloudy. The top of the browning scale was labeled with no browning, and the bottom was labeled with excessive browning. The tops of the color and flavor scales were anchored with highest value, and the bottoms were

anchored with lowest value. The panelists were instructed to make a horizontal mark through the scale and write the corresponding sample number adjacent to the mark. The results were recorded on a scale of one to five. The sum of scores of each sample was divided by the number of panelist and the resulting average expressed as sensory score.

For microbial load the sample was diluted up to 10^{-4} in saline water and then plated on nutrient agar and colony forming units counted.

5. Results

The fruits ripen between February and April. The time from ripening onset to rotting of the fruit was 16±4 days. Skin colour started to change on day 3 after abscission, and a completely yellow colour was obtained by day 12. The sugar content of the berry increases rapidly, acidity (TTA%) decreases, and the pH increases from 2.5 at the green stage of the fruit to 3.32 in fully ripe fruits. Berry skins lose chlorophyll and begin to develop characteristic yellow-gold colour. There was a significant decrease of Vitamin C (Ascorbic acid) due to processing in the order: Sterilize>Blanch/pasteurise>Pasteurise>Blanch. At 22°C the first week shows the greatest decline in vitamin C, which is followed by a period of improved stability. There is a decline in the °Brix value with time in Blanched juice stored at 22°C. At 4°C all treatments show minimal degradation in the °Brix values and Browning index. The sensory score increases as the fruit ripens reaching its peak at °Brix value of 13.5, which corresponds to yellow colour stage of ripening. In response to various storage containers Browning index and vitamin C content decreased in the order of liner>plastic>clear glass>brown glass with time in juice stored at 22°C. There was no significant change in sensory score of Sterilized and Blanched/Pasteurised juice stored at 4°. But at 22°C the decline in sensory score was rapid for all treatments. Measurements of headspace did not affect vitamin C content and browning index when the juice was stored at 4°C but at 22°C they did. Higher processing temperatures reduce microbial flora. Some fruits retained 80% of their vitamin C content after storage at -18°C for two years.

6. Discussion

The pattern of fruit abscission differed among trees, but in most examined individual trees, 80% of the fruits abscised within two weeks which is in agreement with previous authors (Arthachinta, 2000). Ripening of abscised fruits is affected by storage temperatures. After 14 days of storage, fruits kept at 4°C remained green and firm, while those kept at 22°C developed a yellow color and could be squeezed for juice. As ripening progressed they developed a golden yellow color, higher juice content, and lower acidity as also noted for other fruits studied. The fruits began to develop brown spots on the skin one week after the on set of the deeper yellow colour. Similar observations have been made for other fruits studied (Huang and Liu, 2002; Brumell and Harpster, 2001; Mahayothee et al., 2002; Prasanna et al., 2007; Prescott, 2011). These results indicate that climacteric processes in marula start after abscission, which is in agreement with studies on other related fruits (Brumell and Harpster, 2001; Prasanna, et al., 2007; Prescott, 2011). Blanching as treatment of prolonging the shelf life has been used in the canning industries. The effect of this treatment was comparable with previous results from various fruits particularly reduction in vitamin

C content (Brumell and Harpster, 2001; Prasanna, *et al.*, 2007; Rico *et al.*, 2007; Prescott, 2011). Sterilized marula juice remained in a stable state for extended periods but had the greatest loss of vitamin C and sensory score recorded initially which is in agreement with previous studies on these attributes(Brumell and Harpster, 2001; Prasanna *et al.*, 2007; Nassu *et al.*, 2001; Prescott, 2011). °Brix and TTA were not affected by sterilization. Pasteurization least affected the sensory score, the °Brix, and Vitamin C. But the Browning index progressed rapidly under this treatment as previously recorded for other fruits(Brumell and Harpster, 2001; Prasanna *et al.*, 2007; Prescott, 2011). Storage at 4°C showed the highest stability of the juice for all treatments however at 22°C the was a rapid decline of vitamin C and rapid increase in the browning index in the first two weeks which was followed by a more stable quality maintenance for six weeks of the study period. Similar observations have been noted in some fruits previously studied (Brumell and Harpster, 2001; Prasanna *et al.*, 2007; Prescott, 2011). The drop in pH as fruits ripen has been similarly observed for other fruits (Brumell and Harpster, 2001; Prasanna *et al.*, 2007; Prescott, 2011). In order to improve the shelf life of the non-alcoholic marula fruit juice, the appropriate headspace and the most suitable packaging materials should be established and then production of this juice can be embarked-on on a commercial scale. This will make the juice available to children. Thus a wider portion of the population will benefit from this nutritious resource. Efforts to improve on the shelf life of indigenous fruits have been successfully made in Malaysia and other developing countries (Brumell and Harpster, 2001; Prasanna *et al.*, 2007; Prescott, 2011).

7. Concluding remarks

By modifying the physico-chemical conditions marula juice can have its shelf lie extended without compromising its organoleptic properties. The amounts of vitamin C are in agreement with previous studies and this vitamin has been used as measure of stability of fresh fruit juices a factor confirmed by this study. Sugar levels as measured by °Brix and TTA and enzymatic browning are a function ripening and storage conditions and physico-chemical conditions. Temperature in particular is critical for shelf life extension of fresh marula juice.

8. Acknowledgements

This work was funded by the NUST Research Board.

9. References

Adams, M.R. & Moss, M.O. (1995). Food Microbiology Amazon UK 3rd edition.

Aganga, A.A. & Mosase, K.W. (2001). Tannin content, nutritive value and dry matter digestibility of *Lonchocarpus capassa*, *Zizyphus mucronata*, *Sclerocarya birrea*, *Kirkia acuminata* and *Rhus lancea* seeds. Animal Feed Science and Technology 91: 107 – 113.

Allong, R., Wickham, L.D. & Mohammed, M. (2001). The effect of slicing on the rate of respiration ethylene production and ripening of mango fruit *Journal of Food Quality* 24: 13-20.

Arthachinta, C. (2000) Mango research in Thailand. *Proceeding of the Sixth International Symposium on Mango. International Society for Horticultural Science (ISHS)*, 33-35.

Bergh, B.O. & Whitsell, R. H. (1962). A possible dwarfing rootstock for avocados: California Avocado Society Yearbook 46: 55-62.

Blackstone, B. A. (1999). Principles and applications of modified atmosphere packaging of foods Second edition Aspen publishers

Brummell, D. A. & Harpster, M.H. (2001). Cell wall metabolism in fruit softening and quality and its manipulation in transgenic plants. *Plant Mol Biol.* 47(1-2):311-40.

Chao Zhang, Bernhard Trierweiler, Wu Li, Peter Butz, Yong Xu, Corinna E. Rüfer, Yue Ma, & Xiaoyan Zhao (2011). Comparison of thermal, ultraviolet-c, and high pressure treatments on quality parameters of watermelon juice *Food Chemistry* 126:254-260.

Cheikh Ndiaye, Shi-Ying Xu, & Zhang Wang (2009). Steam blanching effect on polyphenoloxidase, peroxidase and colour of mango (*Mangifera indica* L.) slices *Food Chemistry* 113: 92-95.

Cullen P.J., Valdramidis, V.P., Tiwari, B.K., Patil, S., Bourke, P. & O'Donnell, C.P. (2010). Ozone Processing for Food Preservation: An Overview on Fruit Juice Treatments, *Ozone: Science & Engineering*, 32:3, 166-179.

Deliza R., Rosenthal, A. Abadio, F.B.D., Carlos, H., Silva, O. & Castillo, C. (2005). Application of high pressure technology in the fruit juice processing: benefits perceived by consumers *Journal of Food Engineering* 67:241-246.

Dlamini, N.R. & Dube, S. (2008). Studies on the physicochemical, nutritional and microbiological changes during the traditional preparation of Marula wine in Gwanda, Zimbabwe *Nutrition and Food Science*, Vol.38 No.1, pp. 61-69.

Emanuel, P.L., Shackleton, C.M., & Baxter J.S. (2005). Modelling the sustainable harvest of Sclerocarya birrea subsp. caffra fruits in the South African lowveld. *Forest Ecology and Management* 214: 91-103.

Fox, F.W. & Norwood, Y.M.E. (1982). Food from the veld. Edible wild plants of Southern Africa. Johannesburg and Cape Town: Delta 399.

Frank Albert Paine & Heather Y. Paine (1992). A hand book of food packaging Blackie academic and professional an imprint Chapman and Hall London.

Hillman, Z., Mizrahi, Y. & Beit-Yannai, E. (2008). Evaluation of valuable nutrients in selected genotypes of marula (Sclerocarya birrea ssp. caffra). *Scientia Horticulturae* 117 (2008) 321-328.

Liu, X.X. & Huang, K.K. (2002). Textural and sensory properties of α–Amylase treated poi stored at 4°C. *Journal of Food Processing and Preservation*. 26: 35-37.

Indrawati Oey, Martina Lille, Ann Van Loey & Marc Hendrickx, (2008). Effect of high pressure processing on colour, texture and flavour of fruit and vegetable-based food products: a review *Trends in Food Science & Technology* 19:320-328.

Jay, J. M. (1986). Modern Food Microbiology. Van Nostrand Reinhold New York.

Mahayothee, B., Sybille, N., Mühlbauer, W. & Reinhold, C. (2002). Effects of Postharvest Ripening Processes on the Quality of Dried Mango Slices Produced from Thai Mango Cultivars Nam Dokmai and Kaew. *International Symposium Sustaining Food Security and Managing Natural Resources in Southeast Asia - Challenges for the 21st Century* Chiang Mai, Thailand -*January 8-11* :1-3.

Mdluli Kwanele M. (2005). Partial purification and characterisation of polyphenol oxidase and peroxidase from marula fruit (Sclerocarya birrea subsp. Caffra) *Food Chemistry* 92: 311-323.

Michael, John Lewis & Hepell, N.J. (2000). Continuous thermal processing of foods: Pasteurization and UHT sterilization Aspen publishers.

Mizrahi, Y. & Nerd, A. (1996). New crops as a possible solution for the troubled Israeli export market. In: Janick J (Ed). Progress in new crops. ASHS Press, Alexandria, VA. 37-45.

Morga, N.S., Luste, A.O., Tunac, M.M., Balagot, A.H. & Soriano, M.R. (1979). Physico-chemical changes in Philippine *Carabao* mangoes during ripening. *Food Chemistry.* 15:225-234.

Nassu, R.T., Janiice, R., Lima, M.S.A., & Moreira, D.S.F. (2001). Consumers' acceptance of fresh and combined methods for processed melons, mango, and cashew apple. *Bras. Frutic, Jaboticabal – SP* 23:551-554.

Nerd, A. & Mizrahi, Y. (1993). Domestication and introduction of marula (*Sclerocarya birrea* subsp. *Caffra*) as a new crop for the Negev desert of Israel. In: Janick J and JE Simon (Eds). New crops. Wiley, New York. 496-499.

Ojewole, J.A.O., Mawoza, T., Chiwororo, W.D.H. & Owira, P.M.O. (2010). *Sclerocarya birrea* (A. Rich) Hochst. ['Marula'] (Anacardiaceae): A Review of its Phytochemistry, Pharmacology and Toxicology and its Ethnomedicinal Uses. *PHYTOTHERAPY RESEARCH* 24: 633-639.

Palgrave, K.C. (1990). Trees of Southern Africa. Struik Publishers. Cape Town. 959.

Philip E. Nelson (2010). Principles of aseptic processing and packaging, 3rd edition Purdue University presses West Lafayette Indiana.

Prasanna, V., Prabha, T.N., & Tharanathan R. N. (2007). Fruit ripening phenomena--an overview. *Crit Rev Food Sci Nutr.* 47(1):1-19.

Prescott Albert B. (2011). THE CHEMISTRY OF FRUIT-RIPENING. http://en.wikisource.org/wiki/popular science_monthly _volume _12/ february 1878 accessed 20 September 2011.

Rico, D., Martı´n-Diana, A.B., Barat J.M. & Barry-Ryan, C. (2007). Extending and measuring the quality of fresh-cut fruit and vegetables: a review *Trends in Food Science & Technology* 18: 373-386.

Sampedro, F., Rodrigo, D. & Hendrickx M. (2008). Inactivation kinetics of pectin methyl esterase under combined thermal–high pressure treatment in an orange juice–milk beverage *Journal of Food Engineering* 86:133–139.

Shackleton, S.E., Shackleton, C.M., Cunningham, A.B., Lombard, C., Sullivan, C.A. & Netshiluvhi, T.R. (2002). A summary of knowledge on *Sclerocarya birrea* subsp. *caffra* with emphasis on its importance as a non-timber forest product in South and southern Africa. Part 1: Taxonomy, ecology, traditional uses and role in rural livelihoods. *Southern African Forestry Journal,* 194: 27 –41.

Van Wyk, B. & Van Wyk P. (1997) Field guide to trees of South Africa. Struik Publishers. Cape Town. 536.

Van Wyk, B-E., van Oudtshoorn, B. & Gericke, N. (2002). Medicinal plants of South Africa. 2nd edn. Briza: Pretoria 234–235.

Venter, F. & Venter, J. (1996). Making the most of indigenous trees. Briza Publishers. Pretoria : 304.

Von Teichman, I. (1983). Notes on the distribution, morphology, importance and uses of the indigenous Anacardiaceae. The importance and uses of *Sclerocaya birrea* (the marula). *Trees in South Africa;* 35: 2-7.

Permissions

The contributors of this book come from diverse backgrounds, making this book a truly international effort. This book will bring forth new frontiers with its revolutionizing research information and detailed analysis of the nascent developments around the world.

We would like to thank Dr. José C. Jiménez-López, for lending his expertise to make the book truly unique. He has played a crucial role in the development of this book. Without his invaluable contribution this book wouldn't have been possible. He has made vital efforts to compile up to date information on the varied aspects of this subject to make this book a valuable addition to the collection of many professionals and students.

This book was conceptualized with the vision of imparting up-to-date information and advanced data in this field. To ensure the same, a matchless editorial board was set up. Every individual on the board went through rigorous rounds of assessment to prove their worth. After which they invested a large part of their time researching and compiling the most relevant data for our readers. Conferences and sessions were held from time to time between the editorial board and the contributing authors to present the data in the most comprehensible form. The editorial team has worked tirelessly to provide valuable and valid information to help people across the globe.

Every chapter published in this book has been scrutinized by our experts. Their significance has been extensively debated. The topics covered herein carry significant findings which will fuel the growth of the discipline. They may even be implemented as practical applications or may be referred to as a beginning point for another development. Chapters in this book were first published by InTech; hereby published with permission under the Creative Commons Attribution License or equivalent.

The editorial board has been involved in producing this book since its inception. They have spent rigorous hours researching and exploring the diverse topics which have resulted in the successful publishing of this book. They have passed on their knowledge of decades through this book. To expedite this challenging task, the publisher supported the team at every step. A small team of assistant editors was also appointed to further simplify the editing procedure and attain best results for the readers.

Our editorial team has been hand-picked from every corner of the world. Their multi-ethnicity adds dynamic inputs to the discussions which result in innovative outcomes. These outcomes are then further discussed with the researchers and contributors who give their valuable feedback and opinion regarding the same. The feedback is then collaborated with the researches and they are edited in a comprehensive manner to aid the understanding of the subject.

Apart from the editorial board, the designing team has also invested a significant amount of their time in understanding the subject and creating the most relevant covers. They scrutinized every image to scout for the most suitable representation of the subject and create an appropriate cover for the book.

The publishing team has been involved in this book since its early stages. They were actively engaged in every process, be it collecting the data, connecting with the contributors or procuring relevant information. The team has been an ardent support to the editorial, designing and production team. Their endless efforts to recruit the best for this project, has resulted in the accomplishment of this book. They are a veteran in the field of academics and their pool of knowledge is as vast as their experience in printing. Their expertise and guidance has proved useful at every step. Their uncompromising quality standards have made this book an exceptional effort. Their encouragement from time to time has been an inspiration for everyone.

The publisher and the editorial board hope that this book will prove to be a valuable piece of knowledge for researchers, students, practitioners and scholars across the globe.

List of Contributors

Wellman Ribón
Universidad Industrial de Santander, Bucaramanga, Colombia

Malek Shams, Tony Campillo, Céline Lavire, Daniel Muller, Xavier Nesme and Ludovic Vial
Université de Lyon, Ecologie Microbienne Lyon, UMR CNRS 5557, USC INRA 1193, France

Jose C. Jimenez-Lopez
Purdue University, Department of Biological Sciences, College of Science, West Lafayette, IN, USA

María C. Hernandez-Soriano
North Carolina State University, Department of Soil Science, College of Agriculture and Life Sciences, Raleigh, NC, USA

Anthony Pavic
Birling Avian Laboratories, Australia
School of Veterinary Science, University of Sydney, Australia
School of Biotechnology and Biomolecular Sciences, Faculty of Science, University of New South Wales, Australia

Peter J. Groves
School of Veterinary Science, University of Sydney, Australia
Zootechny, Australia

Julian M. Cox
School of Biotechnology and Biomolecular Sciences, Faculty of Science, University of New South Wales, Australia

Veronica Sanda Chedea
Laboratory of Animal Biology, National Research Development Institute for Animal Biology and Nutrition Baloteşti (IBNA), Romania
Faculty of Environmental Protection, University of Oradea, Romania

Simona Ioana Vicaş
Faculty of Environmental Protection, University of Oradea, Romania

Carmen Socaciu
Department of Chemistry and Biochemistry, University of Agricultural Sciences and Veterinary Medicine Cluj-Napoca, Romania

Tsutomu Nagaya, Kazushige Yokota, Kohji Nishimura and Mitsuo Jisaka
Department of Life Science and Biotechnology, Faculty of Life and Environmental Science, Shimane University, Japan

Henry Joseph Oduor Ogola
Department of Life Science and Biotechnology, Faculty of Life and Environmental Science, Shimane University, Japan
School of Agriculture, Food Security and Biodiversity, Bondo University College, Kenya

Zeeshan Feroz
Ziauddin College of Pharmacy, Ziauddin University, Karachi, Pakistan

Rafeeq Alam Khan
Department of Basic Medical Sciences, King Saud Bin Abdul Aziz University of Health Sciences, Jeddah, Kingdom of Saudi Arabia

I.C. Morobe
Walter Sisulu University, Nelson Mandela Drive, Mthatha, Eastern Cape, South Africa
School of Agriculture and Life Sciences, Department of Life and Consumer Sciences, University of South Africa, Pretoria, South Africa
Department of Biological Sciences, University of Botswana, Gaborone, Botswana

M.A. Nyila
School of Agriculture and Life Sciences, Department of Life and Consumer Sciences, University of South Africa, Pretoria, South Africa

C.L. Obi
Walter Sisulu University, Nelson Mandela Drive, Mthatha, Eastern Cape, South Africa
Division of Academic Affairs and Research, Walter Sisulu University, Eastern Cape, South Africa

M.I. Matsheka and B.A. Gashe
Department of Biological Sciences, University of Botswana, Gaborone, Botswana

S. Dube, I. Shereni and T. Sibanda
Department of Applied Biology and Biochemistry, National University of Science and Technology, Ascot, Bulawayo, Zimbabwe

N.R. Dlamini
CSIR Biosciences, Pretoria, South Africa

Printed in the USA
CPSIA information can be obtained
at www.ICGtesting.com
JSHW011417221024
72173JS00004B/560

9 781632 393708